CARPENTRY FOR RESIDENTIAL CONSTRUCTION

Byron W. Maguire

Craftsman Book Company
6058 Corte del Cedro, P. O. Box 6500
Carlsbad, CA 92009

Library of Congress Cataloging-in-Publication Data

Maguire, Byron W., 1931-
    Carpentry for residential construction.

    Includes index.
    1. Carpentry. 2. House construction. 3. Dwellings
——Maintenance and repair. I. Title.
TH5607.M33   1987   694   87-15587
ISBN 0-934041-21-0

CONTENTS

10: Roof Framing   135

11: Shingling   167

12: Cornice   184

13: Window-Unit Installation   204

PREFACE

This book was written for carpenters and builders who need a good reference manual on the carpentry trade. It covers all the carpentry work you're likely to do on homes, apartments and condominiums.

It's not easy to master the carpentry trade. There's a lot to learn. Many skills are required. It's easy to make mistakes. That's why I've tried to keep this book as simple and easy-to-understand as possible — even when covering a complex topic like stair building or roof framing. To make your job easier, I've provided over 200 illustrations. Pictures, drawings, tables and forms should help speed the learning process. Each section includes a list of the tools and materials needed, the estimated manhours required, and a step-by-step guide to each part of the task — from measuring and marking to checking the completed job.

I'll explain both rough carpentry and finish carpentry in detail. The first half of the book covers formwork, framing and sheathing. Even if you're an experienced professional carpenter, this section will show you better ways to form and frame homes. The last two hundred pages of the book are devoted to interior and exterior finish carpentry. There are many good carpenters but a lot fewer good finish carpenters. If your framing skills are better than your finishing skills, the last half of this book will be of special interest to you.

The first five chapters cover important preliminary steps: understanding the plans and specs, estimating manhours and costs, planning how you'll do the work, and the key terms carpenters should understand. True, the information in these chapters isn't strictly part of the carpentry trade. But most carpenters do more than frame homes and install trim and moulding. Many carpenters are also builders and supervisors, estimating costs, reading construction drawings and planning how the work will be done. If your job requires that you handle these important tasks, you'll appreciate the information in the first five chapters.

All carpentry work explained in this book is broken down into simple steps or procedures that are easy to follow and understand. These procedures come after the text at the end of each chapter. The text provides a word picture of the typical job — from beginning to end — and includes cautions and tips to prevent mistakes that can cause injury or result in the waste of time or material. The procedures show you step-by-step how to do the work.

Each procedure is identified with code letters and a number that should make it easier for you to find the information you need. For example, the first basic framing routine is coded BF1. The B stands for *basic building,* F for *framing* and 1 for the first of many routines.

Because some procedures refer to other procedures by code number, you need a convenient way to find the section you're looking for. There are two ways to do this. First you can use the table of contents. Procedure codes are printed in bold face type to make them stand out better. The second way is to use the reference list that begins on page 382.

Whether you're a skilled professional carpenter or an apprentice just learning the trade, this manual should help you discover better, quicker ways to do professional-quality carpentry on any residential job.

Finally, let me admit that no good book on carpentry could be written without information from many sources. I've received assistance from many companies and individuals and want to acknowledge them before we get into the first chapter. Contributions from Armstrong World Industries, Masonite Corporation, Stanley Tools, and United States Gypsum Company are most sincerely appreciated. I'll also offer a special thanks to my wife, Betty, for her secretarial and proofreading assistance.

Byron W. Maguire

UNIT ONE

Planning for Carpentry and Building

The five chapters of Unit One provide you with a simplified method of planning any carpentry job that you might undertake. Chapter 1 encourages you to make decisions that will get you started. It discusses various types of planning and introduces you to the organization of the book. Chapter 2, Drawings and Specifications, and Chapter 3, Estimating, provide insight into tasks and describe methods of incorporating ideas into the plan for the job. Chapter 4, Programming, provides a set of forms that should be used for each project. When you have completed the forms you will thoroughly understand the magnitude of an undertaking, its total cost in materials and tools, and the time that will be required—*before* you do any work or buy any stock. Chapter 5 leads us further into the world of carpentry and building through terminology.

1

PLANNING

Usually a pressing need elicits an immediate response. Once the need is satisfied, we often wish another object had been purchased or another method used to obtain the object. Too often the solution selected results in considerable waste. Planning could avoid much of this and is, therefore, extremely important.

ELEMENTS OF PLANNING

Planning means simply that a process of preselection is being used. Alternatives are identified, examined, filed, and discarded; and finally the best one is chosen. To make this choice, the alternatives must be listed and studied, and their details, characteristics, and usefulness must be included. They must also be categorized as temporary or long-range solutions.

As the objective of this book is to provide you with the tools for improving and maintaining residential homes, shops, and offices, most of our suggestions will be based upon the long-range or permanent solution rather than the temporary one. This approach is most desirable— if for no other reason, because it eliminates the need to redo the task. On the other hand, a temporary solution is often essential to the health, comfort, or preservation of property and family. In these cases, for example replacing a broken front-door hinge, the task cannot be put off (Figure 1-1). Fortunately, most of the tasks in this category can be performed quickly; and later, as time permits, a more thorough job can be done. Let's expand this idea a bit.

The hinge on the front door pulled away from the jamb, and the door does not open or close correctly. As a temporary remedy, longer screws could be used to attach the hinge. As you will see in Chapter 15, this method is not adequate for a permanent repair but will provide a temporary cure. The long-range cure will require drilling the holes (old screw holes), filling them with glue, then inserting short dowel pins. After the glue dries, the dowels are clipped even with the mortised hinge surface. Finally, new screws of the correct size are used to reinstall the hinge.

Figure 1-1 Broken Door Hinge

So, planning explores the choices of temporary or permanent cures for problems. How does planning apply to materials? You must plan the use of materials carefully so that waste will be held to a minimum. For instance, if structural members such as 2 x 4s or 2 x 6s are to be used, each should be ordered according to its intended use. Studs should be bought in 8-ft lengths and ceiling joists for a 12-ft span in 12-ft. lengths.

BUILDING PLANS

Without a blueprint (plan), a material list cannot be determined accurately. The plan, whether it be a formal architectural drawing, a simplified drawing, or a detailed written study, must relate the job in all its detail. It must take into account existing circumstances, detailed tasks, and specifications that must be met. Some of these data are not readily apparent, because experienced people usually read plans and interpret them in terms of the tasks to be accomplished. Units II and III translate these tasks from plans into simple procedural steps.

If your work is to be rather extensive—adding a porch or changing the walls within your home—then perhaps you should seek professional help in making the floor plans, elevation plans, and detail drawings. In the long run this cost may be repaid several times over. Such things as structural collapse and electrical or plumbing problems will be identified and avoided, saving much grief and considerable money. Another consideration is the *building code*. Since your property is probably insured, a violation of the code may put you in a difficult position.

On the other hand, simple plans may be all you need. For example, you may want to panel a room after installing a new closet. You can easily draw this plan by using the ideas examined in Chapter 2. Be cautious in making your decision; many factors will become evident, and you must weigh each one. Make decisions as to whether to buy plans or draw them yourself on a job-by-job basis.

DEFINING THE ROUTINES TO BE USED

You have your plan—written and/or drawn—your selection of tasks, *routines* that when accomplished in a particular sequence will result in a completed job. For example, let's examine the task of cutting a rectangle in a piece of wall paneling for an electrical outlet. In Chapter 21 panel installation is discussed, and routine IP5, *Cutting out for Windows, Doors, and Utilities,* details the steps needed to accomplish the task. Briefly the sequence would go as follows:

1. Determine the top of the panel.
2. From a reference point (such as the edge of the previous panel), measure and record the distances to both sides of the outlet box on the panel.
3. From the ceiling down, repeat step 2.
4. Draw a rectangle that connects all four marks.
5. Predrill two holes, in opposite corners, with a spade bit 3/8 in. in size.
6. Cut out a rectangle with a keyhole or saber saw.

As you see, the steps detail the task. The object of your planning is to decide if the task is needed and how many times it will be used.

SCHEDULING THE WORK

Planning the work effort requires accumulating specific details within each task and eventually totaling these in groups. There are three primary areas of interest: (1) time, (2) material, and (3) special considerations. For each routine in Units II and III, an outline of estimated manhours and number of workers needed is included, together with a bill of materials from which total requirements can be calculated. Finally, from your personal knowledge and from the details of the design, special considerations can be listed. These three items are the keys to sound planning for three reasons: (1) ordering materials at the proper time keeps labor and cost to a minimum; (2) estimating manhours alerts you ahead of time to the need for a helper and shows you clearly how much work can be accomplished within a given time frame; and (3) a sequence of tasks will ensure a logical progression from start to finish.

2

DRAWINGS AND SPECIFICATIONS

The data in this chapter provide you with insight into the various details that make up drawings and specifications. From these data you can determine whether the services of an architect and draftsman are needed or whether you can draw the necessary plans yourself. They will also enable you to understand the various parts of a set of plans.

The various views usually provided in a plan are the *foundation*, the *floor plan*, the *four outside elevations*, the *electrical*, the *heating and plumbing*, and the *detailed sectional views*. This chapter briefly explains each; in addition, it provides some suggestions for simplified drawings and introduces you to the standard symbols used in architectural drawings.

SCOPE OF THE WORK

Whether you draw a plan in simplified or extremely detailed fashion, you must determine the views you will need. Consider these examples:

1. Making a laundry room from part of the garage: Two partitions will probably be needed, one of them with a door. A simple floor plan should be made, including the location and details of the new walls. An elevation plan should be made for the wall with the door. To aid you in making these drawings, refer to the illustrations of various jobs in subsequent chapters, in this case Chapter 7.

2. Pouring a new concrete patio: A foundation-type floor plan should be drawn. This plan should provide size and positioning data for the patio. Either a detail or an elevation plan should be drawn to show footings, depth of concrete, slope of concrete, and reference points.

3. Adding a room on the house: In this type of job, unless you have considerable skill, a professional draftsman should draw up the plans. Included should be all the types of drawings listed earlier.

These three examples provide some insight into when plans should be drawn. So that we understand what the various plans represent, study the following examples.

DESCRIBING THE PARTS OF A STRUCTURE

Five parts of a structure are examined: the foundation plan, the floor plan, the elevations, the detail drawings, and a perspective view.

Foundation Plans

Figure 2-1 shows a typical *foundation plan*. It contains the following specification data:

1. The outside-perimeter dimensions.
2. The intermediate dimensions, as for positioning center pilasters and piers.

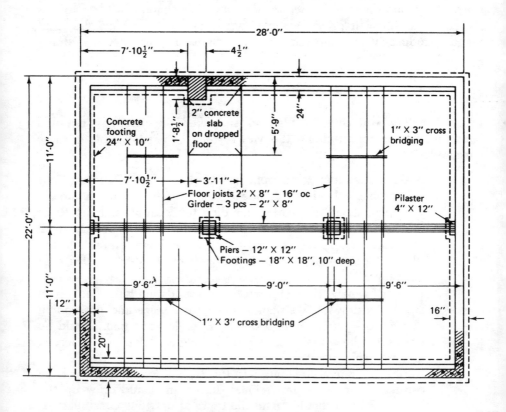

Figure 2-1 Foundation Plan

3. The width of the wall: 12 in. and 16 in.
4. The concrete footing dimensions: 24 in. x 10 in. x 24 in. deep.

A foundation plan is a plan view of a structure projected on a horizontal plane that passes through at the level of the top of the foundation. It contains all the details necessary to plan the work effort, estimate quantities of materials, and lay out the work. This particular plan defines a 12-in. block as the size to be used for the wall. Note that the wall is to be centered on the 24-in. footing. Also note that the piers are 12 in. x 12 in. with footings at 18 in. x 18 in. x 10 in. deep, how the floor joists will run, and that they are spaced on 16-in. centers.

Floor Plans

The *floor plan* (also called a *building plan*) is developed as shown in Figure 2-2. You can see that the floor plan basically represents a horizontal cross section of the walls of a room or house. A floor plan includes the following information:

1. The length, thickness, and character of the building walls on that floor.
2. The widths and locations of the door and the window openings.

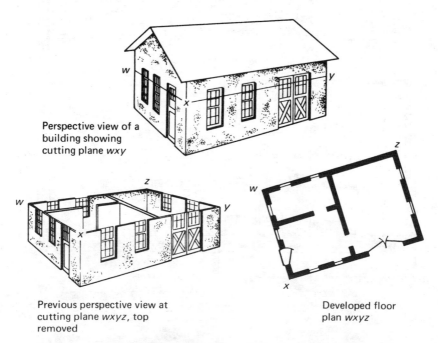

Perspective view of a building showing cutting plane *wxy*

Previous perspective view at cutting plane *wxyz*, top removed

Developed floor plan *wxyz*

Figure 2-2 Floor Plan, Development

3. The length and character of partitions.
4. The number and arrangement of rooms.
5. The type and location of utilities.

Figure 2-3 shows each of these items in detail. It will be helpful for you to correlate the details shown in the illustration with the following list:

1. The total length of the wall, intermediate dimensions, and widths and locations of window and door openings.

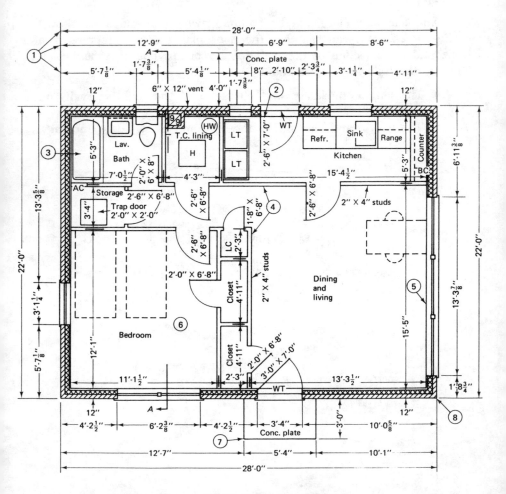

Figure 2-3 Floor Plan, Detailed

2. The door size and direction of opening.
3. The symbols for the bathroom appliances.
4. The interior-stud-wall placement with door openings.
5. The mullion-type windows.
6. The room arrangement.
7. The concrete platform (also on back door).
8. The concrete block wall with brick veneer.

Figure 2-4 reveals the heating layout for this house, which calls for hot-water heating. In installations of forced-air systems, the duct work would be drawn with register openings and pipe runs identified.

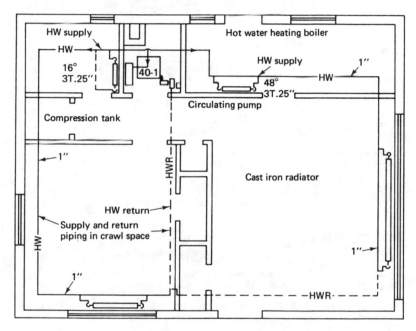

Figure 2-4 Heating Layout

Figure 2-5 illustrates the electrical outlets, switches, and receptacles for the house. Note the specific symbol used for each type of unit.

Elevations

The next type of drawings that may be needed is the *elevation plan*. A view of each wall should be made, showing the special details of that side of the wall. The drawings will show the height of walls, windows, and doors and the roof line and its slope or pitch.

Figure 2-6 is an elevation plan and the details are as follows:

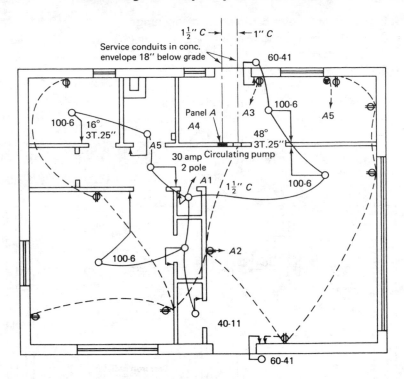

Figure 2-5 Electrical Layout

1. The window height from the finished floor level.
2. The wall height from the finished floor level.
3. The roof pitch: 4 in. and 12 in.
4. The wood trim for the overhang and eave.
5. The height of the chimney above the ridge of the roof.
6. The type of shingles on the roof.
7. The door height.
8. The flashing.
9. The vent in the gable end.
10. The brick veneer.

An adaptation of the elevation drawing is one that details the cabinetry in a room. This type of drawing will include the height, length, position, and size of cabinets, and whether they are base units, wall units, or bookcases. For this type of use, an elevation drawing is essential.

Not all information can be provided in the elevation and floor plan, so another type of drawing is made, a detail drawing.

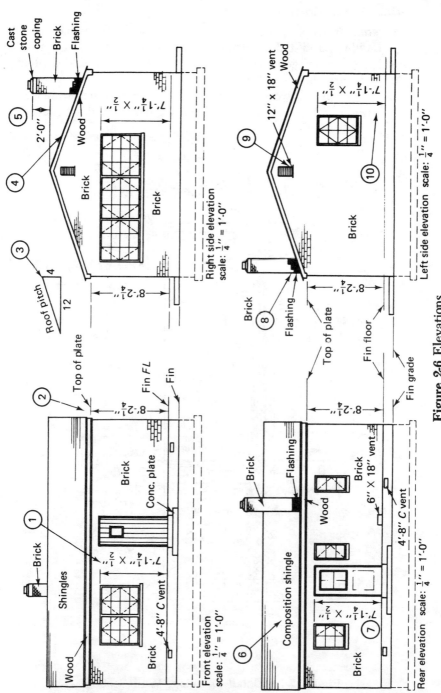

Figure 2-6 Elevations

Detail Drawings

A *detail drawing* is prepared on a larger scale than that used for general drawings, and it shows features that do not appear, or appear on too small a scale, on general drawings. The wall sections described in the elevations are shown in full detail.

Figure 2-7 shows two wall sections: A, a cement block wall section, and B, a wood frame wall section. Details are included whenever the

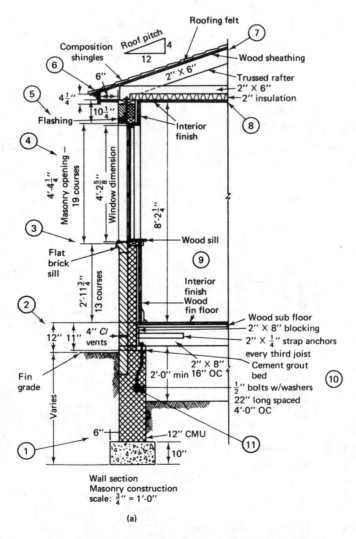

(a)

Figure 2-7A Detail Drawing, Wall Section

information given in the floor plan or elevations is not sufficient for the craftsmen working on the job. Note the various details in Figure 2-7A:

1. The footing and foundation data.
2. The vents.
3. The finished window-sill height.
4. The window dimensions.
5. The flashing over the windows.
6. The overhang and cornice detail.

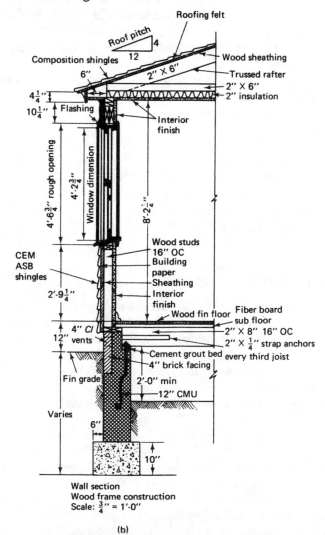

Wall section
Wood frame construction
Scale: $\frac{3}{4}'' = 1'-0''$

(b)

Figure 2-7B Detail Drawing, Wall Section

7. The roof composition.
8. The insulation.
9. The interior trim and finish.
10. The floor-joist detail.
11. The anchor-bolt detail.

Data can be identified in Figure 2-7B in a similar manner. Note the differences, in the use of studs, asbestos siding (ASB), and a header over the window.

Figure 2-8 shows additional applications of detail. In this illustration, A illustrates the manner in which the outside door jamb is installed and trimmed and how the siding butts against it. B illustrates the head over the door, its trim, and the rain (drip) cap. C illustrates the details for the window sill, jamb, and trim, and the window head, jamb, casing, and drip cap. D provides detail for building the overhang into a box cornice. Note that quite a few pieces of stock are required to complete the box cornice.

In the preparation of any drawing that deals with subjects previously identified, refer to the chapter which discusses that phase of the project for help in defining the details needed.

What has been presented in this chapter so far has been an introduction to blueprint reading for residential home plans. The balance of the chapter provides suggestions and examples from which you can successfully draw simple plans to meet most of your requirements.

LAYING OUT THE WORK

The best way to draw a plan that will represent the work to be done is to draw to a prescribed scale. This will place each element of the drawing in its proper relationship. As each element is drawn, its detail is converted to the scale that you have selected. Once all details have been entered on the plan and the various profiles drawn, you will collect the specifications that are to accompany the plan. Let's examine these two elements: scaling a plan and listing specifications.

SCALING A PLAN

Refer back to Figure 2-6 for a moment and locate the *scale* notation under each elevation. It is printed *SCALE: 1/4" = 1' − 0"*. This means that each 1/4 in. on the drawing is equal to 1 ft of actual building size. Any portion of an inch or multiple of an inch may be selected as the scale. If you anticipate continued drawing of plans, it would be useful for you to have an *architects' scale* similar to the one shown in Figure 2-9. Note that on the scale shown, the 1/4-in. and 1/8-in. scales occupy the

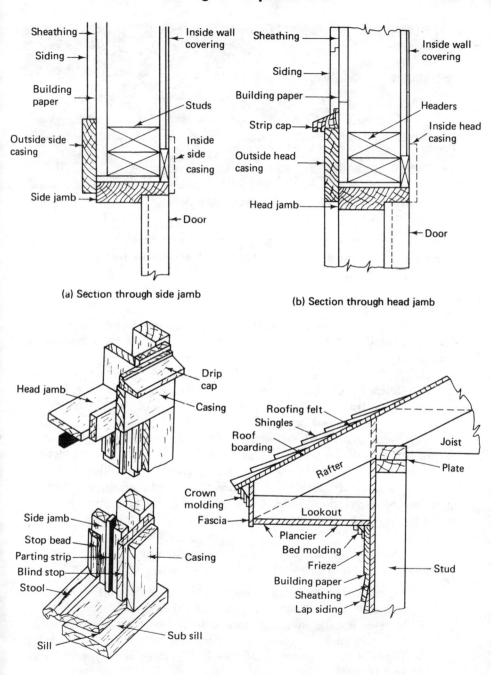

(a) Section through side jamb

(b) Section through head jamb

(c) Upper-lower corner details double-hung window frame

(d) Closed or box cornice

Figure 2-8 Detail Drawing, Door, Window, and Cornice

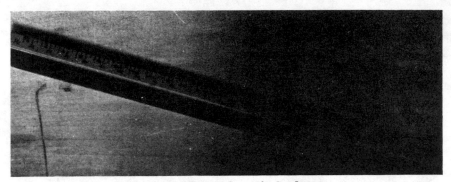

Figure 2-9 Architect's Scale

same plane but are at opposite ends. Note also the upper 1-in. and 1/2-in. scales. The divisions at the ends represent fractions of feet.

SPECIFICATIONS

A considerable amount of information can be represented in the details of plans, but the builder must have more data than can be included on the drawings. Information of this type includes criteria for materials, such as grade and type of plywood, and standards of workmanship. Such information is presented in lists known as *specifications* ("specs" in the trade). There are usually two types of specifications: (1) general, and (2) special or specific.

A *general specification* contains data such as "All wood work shall be protected from the weather; nailing will be done as far as practical, in concealed places; lumber will be sanded four sides; millwork will be sanded smooth;" etc. Specific *specifications* are usually grouped in sections or under headings that describe the phase of the job. Such headings might include "(1) Heating, (2) Plumbing, (3) Framing, (4) Windows, (5) Doors," etc.

For your projects the specific specification list is extremely important, and it can best be collected from a variety of sources:

1. Routines that you plan to use; these will give you the estimated material: its size, style, types, and quantity.
2. Lists of installed units that need to be matched to other rooms or areas, such as doors, trim, paneling, windows, and flooring.
3. Types of construction details, such as rafters, roof type, shingles, and framing.

These specifications can be listed on a separate sheet of your plan or be entered on the material listing worksheet of the programmed plan described in Chapter 4.

The general specifications can be minimal, because you will be doing the work. The specifications may be entered on page 1 of the programmed plan as the job description, thereby completing the first entry of the plan.

We have discussed a variety of facts about plans, their detail, and their usefulness, as well as information on how to specify elements of the plan that cannot easily be drawn. Figures 2-10 through 2-13 will aid you with your drawing.

Figure 2-10 Drawing Kit

Outlet		Special purpose outlet	
Pull switch		Single pole switch	S
Duplex convenience outlet		Three way switch	S_3
Range outlet		Power panel	

Figure 2-11 Electrical Symbols

Door symbols

Type	Symbol

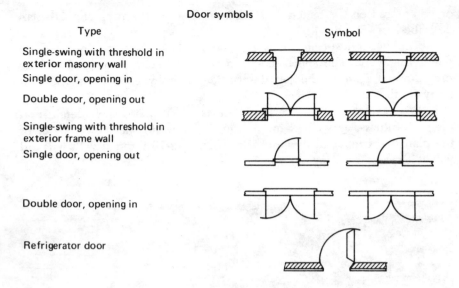

Single-swing with threshold in exterior masonry wall
Single door, opening in

Double door, opening out

Single-swing with threshold in exterior frame wall
Single door, opening out

Double door, opening in

Refrigerator door

Window symbols

Type	Wood or metal sash in frame wall	Metal sash in masonry wall	Wood sash in masonry wall

Symbol

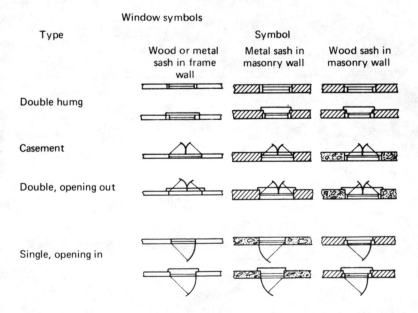

Double hung

Casement

Double, opening out

Single, opening in

Figure 2-12 Architectural Symbols

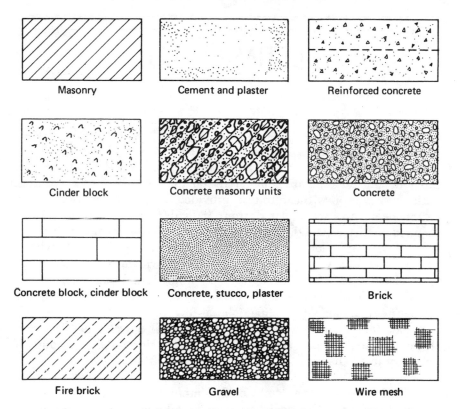

Figure 2-13 Masonry Symbols

3

ESTIMATING

Estimating must be used as a method to provide a reasonable, accurate listing of the resources that will be needed to fulfill a planned job. Material, labor, and tool requirements must all be included. This chapter provides aids to estimating materials on an individual task (routine) basis, and interprets the formulas provided within the routines. Labor requirements are accumulated from individual task routines and are summed to determine the total requirement. Then the individual task requirements are plotted along a time-line-plan, which details when a phase of work is scheduled and how long it will take to perform. All these elements will form a part of your programmed plan (Chapter 4).

Recall from Chapter 1 the example of the broken door hinge. In estimating that simple, routine job, a few basic materials would be required. Your estimate would list: (1) glue, (2) dowel, and (3) screws. It would also list approximately 1 hour to perform the task, and the necessary tools (hammer, drill, 1/4-in. drill bit, 3/4-in. chisel, and screwdriver).

Estimating this task is obviously simple, because it is a single task and relatively easy. Remodeling a room or adding a patio, however, are extensive tasks that encompass many routines.

ESTIMATING MATERIAL REQUIREMENTS

Materials must be estimated based upon their individual unit measurements: for instance, nails per pound; screws per unit; lumber per lineal foot, board foot, or square foot; cement blocks per unit; concrete per cubic yard. Numerous tables and charts are provided in this chapter and in subsequent chapters associated with routines that require their use. Let's start with measuring lumber.

Board-Foot Measurements

A board foot is equal to a piece of stock whose measurements always equal 12 in. x 12 in. x 1 in., where 12 in. equals width and length and 1 in. equals thickness. Stock lumber of all kinds is bought and sold in lots of 1,000 board feet. This means that a quotation of $50 per 1,000 bd ft is equivalent to $0.05 per lineal foot ($50 ÷ 1,000 bd ft). If stock is classified as 1 in., its finished or dressed state equals 3/4 in. The 1/4 in. of waste results from dressing (planeing smooth) both surfaces. Even though you only get a board 3/4 in. thick, you pay at the rate charged for a 1-in. board. To use a 2 x 4 as an example, even though the board is actually only 1-1/2 in. thick, you pay as if it were a 2-in.-thick board.

The examples below will aid you in deriving the board-foot measurement of various sizes of stock lumber.

Formulas are included that give you the factor necessary to convert the size of stock to board feet (or use Table 3-1). Let's use as examples 1-in. stock and 2-in. stock. Keep in mind that the formula for board-foot measurement is a piece of stock equal to 12 in. x 12 in. x 1 in.

TABLE 3-1: Factors for Board-Foot Measurement

Nominal Thickness (in.)	Width (in.)				
	4	6	8	10	12
1	1/3	1/2	2/3	5/6	1
2	2/3	1	1-1/3	1-2/3	2

Formula: factor × length of stock = board feet. Select a factor by using a thickness and an appropriate width: e.g., 1 x 4 = factor of 1/3; multiply factor (1/3) × length of board (12 ft) = 4 bd ft.

Example A: 1 in. x 4 in. x 10 ft piece

Take the second dimension (the width: 4 in.). It is equal to one-third of 12 in.; therefore, the board-foot factor is 1/3. Multiply the factor by the length and by the thickness to obtain the board-foot measurement:

1 in. × 4 in. × 10 ft =
1 in. × 4 in./12 in. × 10 ft =
1 in. 1/3 × 10 ft = 3-1/3 bd ft

Or, stated another way:

1. Multiply the thickness × the width to obtain the factor.
2. Multiply the factor × the length to obtain the board feet.

Thus:

1. 1 in. × 4 in. = 1/3 board-foot width (1/3 of 12 in.).
2. 1/3 × 10 ft = 3-1/3 bd ft.

Therefore, 3-1/3 bd ft are equal to 1 in. × 4 in. × 10 ft.

Example B: 2 in. x 8 in. x 8 ft piece
2 in. × 8 in. = 16 in. = 1 ft 4 in., *or* 1-1/3 bd ft per 12 in. of length.
1-1/3 × 8 ft = 10-2/3 bd ft.

(*Note:* Fractional portions of a board-foot measurement may be translated into decimal equivalents or left as fractions. However, most people in the lumber industry express lengths and quantities in fractional parts of a foot.)

Square-Foot Measurements

Square-foot measurements are based upon a flat surface 12 in. × 12 in., whose thickness *usually* does not exceed 1 in. These dimensions apply, of course, to all varieties of paneling: from sheathing plywood to wallboard to floor tile. In every case, sheet goods are ordered by specific sizes *or* by the square foot. Table 3-2 lists the standard sizes of sheet goods available in lumberyards and most frequently used in home construction. The most common size is the 4 ft x 8 ft sheet. This stock size represents 32 sq ft.

TABLE 3-2: Square Feet in Standard Sheet Stock

Sheet Goods		No. Sheets					
Width (ft)	Length (ft)	1	2	3	4	5	10
1	8	8	16	24	32	40	80
2	8	16	32	48	64	80	160
4	4	16	32	48	63	80	160
4	8	32	64	96	128	160	320
4	10	40	80	120	160	200	400
4	12	48	96	128	192	240	480
4	16	64	128	192	264	320	640

Formula: 1 sq ft = 12 in. wide × 12 in. long × 1 in. thick.

Estimating the Number of Pieces Needed

To estimate the number of structural members required for a part of the job, the *routine* defining the job will have an entry or entries similar to the following examples. These extracts and their explanation are representative of all material requirements that must be individually calculated.

Example A: (Extract from Routine BC3, Forms for Footings):
18 to 24 in. of 1 x 3 for ——————— stakes at 4-ft intervals = ———————
                                    no.   lineal
ft of 1 x 3 plus 1 stake for each corner [Where no. × length (18 or 24 in.)
= lineal feet].

Recommended industry procedure is to stake a footing form each
4 ft of the length of the footing (plus 1 stake for each corner). The
stakes should be cut to a length equal to 3 to 4 times the height of the
footing. The type of ground that the stake is to be driven into will
dictate the length of the stake.

1. Select a length of stake that will adequately do the job—18 to
   24 in. or, in some cases, longer or shorter.
2. From your foundation plan, compute the total perimeter length
   and multiply by 2 for total lineal feet of footing form to be installed.
3. Divide the total lineal feet by 4 (feet) to obtain the number of
   stakes needed. Add one stake for each inside and outside corner
   (i.e., 4 corners need 4 additional stakes).
4. Multiply the number of stakes needed by their individual length
   and list the results in lineal feet of 1 x 3 needed.

In most applications this type of entry will require that you use
your plan and/or specification to aid in developing the requirements
for materials.

Example B (Extract from Routine BF3, Constructing a Window-
Frame Unit):

Two 2 x 6s, 2 x 8s, 2 x 10s, or 2 x 12s x length (L) of header. (See
Tables BF3A and BF3B.)

TABLE BF3A: Headers for Non-Load-Bearing Walls

Span (ft)	Lumber Size
Less than 3	Two 2 x 4s on edge
3–5	Two 2 x 6s on edge
5–8	Two 2 x 8s on edge
8–12	Two 2 x 10s on edge
Over 12	Two 2 x 12s on edge

TABLE BF3B: Headers for Load-Bearing Walls

Span (ft)	Lumber Size
Less than 3	Two 2 x 6s on edge
4–6	Two 2 x 8s on edge
6–10	Two 2 x 10s on edge
10–16	Two 2 x 12s on edge

A header member that is installed over a window opening must be designed to carry a load. The load may be joists and rafters or just the wall above the window. If the wall supports joists and/or rafters, the wall is *load-bearing* and Table BF3B would be used. If the wall does not support rafters or ceiling joists, it is non-bearing and Table BF3A would be used. The wider the span of the header, the greater the total load placed upon it and therefore the wider the stock needed.

1. Two structural members are needed to form one header (Figure 3-1). Note that the two members are nailed together (with spacers) and used by standing them on their edge (the narrow edge).

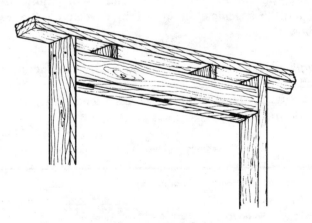

Figure 3-1 Header

2. Next you must determine, from your plan, the length needed. There will be different lengths for various windows and doors. Make a list of each *different* length.

3. With the lengths listed, identify whether the wall is load-bearing or non-load-bearing and indicate this following the header's length.
4. Use Table BF3A or BF3B to determine the proper size of stock member to use. For example: the front door is 36 in. wide and is a load-bearing wall. Table BF3B is used and shows that the minimum-size member to use is a 2 x 6. So your estimate will list: 2 x 6 x 42 in. headers or one 2 x 6 x 7 ft.

ESTIMATING LABOR REQUIREMENTS

Following is a method by which you can program your work schedule from your completed program plan. The method is called the *time-line plan*. Its primary objective is to provide a visual plan that combines times and occurrences. Although it is extremely easy to make, there are some decisions to be made and some simple mathematical problems to solve.

Time-Line Plan

Make a layout similar to the one in Figure 3-2. It should include the entries shown on the left. They are the date line, major step line, material requirements line, and minor step line. Each major and minor step should be assigned a completion date. The materials must be assigned dates so that all or part of the material will be delivered beforehand or on the day needed. Beneath each major step, minor step, and material step, briefly state what is needed. Your previously completed programmed plan will provide this information. You will need to use all sections of the programmed plan when making the time-line plan.

You may have noted that this time-line plan is based upon the availability of a 4- to 5-hour span of time on successive Saturdays. Therefore, each projected date contains a Saturday calendar date and the estimated manhours needed to perform the task or tasks described.

From your programmed plan the major steps can be taken from the task-sequence list; the description would provide additional detail. These major tasks would be listed along the major step line in the order in which you expect to perform them. The materials needed for each task should be taken as a whole from the summary sheet or in part from the individual requirement list, by routine. Which of these to do will depend on individual circumstances. Controlling factors would include weather, storage capability, and financial considerations.

The minor steps can also be taken from the programmed plan. They might be shown as preliminary steps, such as cutting studs, making

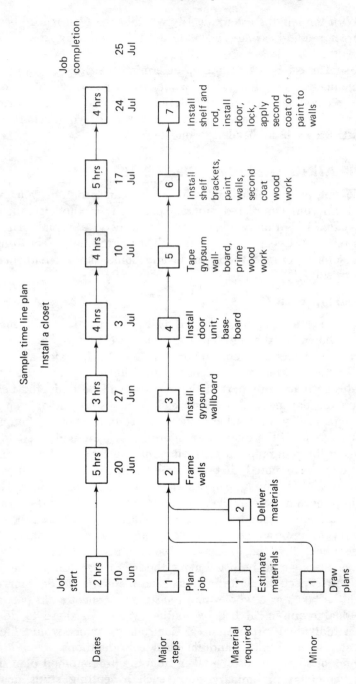

Figure 3-2 Sample Time-Line Plan

outside corners, removing materials prior to new installations, shoring ceilings, prefinishing stock, and other "getting-ready" steps.

The time-line plan is very useful. It is easy to make and provides a quick reference. It outlines in illustrative form the data accumulated in the programmed plan.

4

PROGRAMMING

Programming any job requires some time and effort. It includes planning so that no waste results and so that job completion is satisfactory. In building and home repair, five major objectives must be outlined when programming a job:

1. Defining the job.
2. Listing the tasks required.
3. Listing the materials required.
4. Listing the tools required.
5. Scheduling and time allotment.

The programmed plan that follows incorporates all the required parts, and an explanation of each is provided in the following pages.

DEFINING THE JOB

Select the specific *title* that most closely describes the job to be performed. The title should include sufficient description to specifically define the scope of the work to be done. Such a title as *Install a New Front Door in the House* would be appropriate.

The next phase in the plan is to describe the work in terms or details that generally identify its scope. (Remember: an outline or the specifications usually describe the job adequately.)

SAMPLE PROGRAMMED TITLE PAGE

TITLE: INSTALL A NEW FRONT DOOR IN THE HOUSE

Method A: Description text

Description: Purchase a 3 ft 0 in. x 6 ft 8 in. x 1-3/4 in. 8-light colonial door with 3 lights in the top panel. Use the same lock and hinges as were used in the old door. Buy new weatherstripping to be installed on the door frame. Replace the aluminum threshold (36 in.). After fitting the door, transfer the hinge placement to the new door. Measure and install the old lock into the new door in the same place as it was in the old door. Make a sketch of the lock's placement.

Method B: Specification text
Description: 1. Purchase a 3 ft 0 in. x 6 ft 8 in. x 1-3/4 in. 8-light colonial door with 3 lights in the top panel.
2. Use the same hinges and locks.
3. Buy new weatherstripping for the jamb.
4. Replace the aluminum threshold.
5. Fit the door and mark the placement of hinges and lock.
6. Install the door, lock, and threshold.

If possible, list and describe these elements in the order in which you expect to install them. If you are not accurate, it really doesn't matter. What is significant is that you are aware of the details and have listed each part of the total job as well as you can. Include in your description one other element—the type of drawing that is to be used.

LISTING THE TASKS REQUIRED

Form PP100A is used when listing the tasks. Select those tasks (routines) that will be needed to do the job from the list of routines in Appendix A. This will direct your attention to the routines and the specific chapter that contains the details of each routine. The description previously written, together with the plan and its specifications, provides the guide to the routines needed.

On page 1 of the programmed plan you will list each routine by its number, title, estimated manhours, number of times used, and chapter number. The entries will be similar to those shown in Figure 4-1.

LISTING THE MATERIALS REQUIRED

In order to compile a materials list from which an estimate can be obtained, a two-step process is detailed in the programmed plan. The parts of the plan are: (1) transcribing the materials needed from the routine's material list to a form, and converting stock lumber into board feet if required, and (2) aggregating the materials according to category and type onto a summary materials list form. Form PP101A of the program plan provides space for transcribing the material requirements from the various routines that will be used. This form also contains the

Routines to be used:

Routine no.	Chap. no.	Routine title	Est. no. of men	No. times used	Ext. M/H
RRD 2	21	SETTING A HINGE AND HANGING			
		THE DOOR	1	1	1.0
BD 1	14	EXTERIOR DOOR UNIT INSTALL-			
		ATION	1	1	3.5
RRD 3	21	SETTING A LOCK	1	1	45 min
RRD 5	21	FITTING A THRESHOLD	1	1	1.0

Form PP100A

Figure 4-1 Routines To Be Used (Part of Programmed Plan,
Form PP100A)

factoring data needed for stock lumber conversion from lineal feet to board feet. (The way to use the factors was discussed in Chapter 3.)

Select a routine from page 1 of the programmed plan. List its number in the block labeled Routine No., as shown on Figure 4-2. From the material list in the book, copy the types and sizes of materials needed. After each line entry, list the total quantities needed. For instance, Routine BF3, Constructing a Window-Frame Unit, is assumed to be used three times, see Figure 4-2. This quantity factor could easily be taken from page 1 of the plan. Under the *quantity/name/dimension* column, each line contains an entry taken from Routine BF3. The first entry lists *two 2 x 4 x 8 ft common studs*. Assuming that three BF3 units are being built, the total quantity of common studs needed is six. This amount is listed in the *total column*. Each succeeding line is treated in the same manner. All varieties of material are included in the listing. Each item will be categorized on the summary sheet.

The second part of the initial task is to convert stock lumber to board-foot measurements. This task is not always required; as explained in Chapter 3, selected lengths of lumber frequently result in almost no waste, whereas ordering in board feet may prove costly.

The final phase of listing the materials required is to accumulate the variety of stock lumber, sheet goods, and hardware. Each size or

Job material requirement list

Materials:

Routine no. ___**RRD 5**___

Quality/name/dimensions	Total req.	BD ft
36" ALUMINIUM THRESHOLD	1	
1¾" x #10 FH WOOD SCREWS	6	
CAULKING COMPOUND	1 TUBE	
30 LB FELT PAPER	3 SQ.FT	
GLAZING COMPOUND	½ PT.	
#8d FINISH GALVANIZED NAILS	6	

Materials:

Routine no. ___**BF 3**___

Quality/name/dimensions	Total req.	BD ft
2 – 2 x 4 x 8' COMMON STUDS	6	32
32 LIN Ft 2x4 JACK STUDS	96 LIN Ft	64
8 LIN Ft 2x6 HEADER	24 LIN Ft	24
12d COMM NAILS	3 LB	

Form PP101A

Figure 4-2 Listing Material Requirements by Routine

type of material should be listed separately in the summary sheet (Form PP102A). There is a simple process that should be used to make the list. Start with the first material requirement from Form PP101A and list it on the first line entry of the summary. Assume the entry to be a listing of 2 x 4s. It would look as shown in Figure 4-3. The next entry could come from the same routine if more 2 x 4s were required. Let's assume that there is another 2 x 4 requirement in the routine; its entry into the summary would look as in line 2 of the partial summary form (Figure 4-3). The second routine could have 2 x 4, 2 x 6, plywood, and 1 x 8 requirements, and they would be listed, for instance, as lines 3, 4, 5, and 6.

The right side of the material summary list has been prepared with less room for entries. Notice that the heading shows "Hardware." In this column you can expect to list nails and screw requirements, and roll goods such as tarpaper or screen wire.

Material summary list

Stock	Unit price	Total price	Hardware	Unit price	Total price
2 – 2 × 4 × 16'					
3 – 2 × 4 × 12'					
16 – 2 × 4 × 8'					
4 – 2 × 6 × 12'					
1 – 4 × 8 × ½" PLYWOOD GRADE CD					
10 – 1 × 8 × 16' No2 COM PINE					
Total estimate					

Form PP102A

Figure 4-3 Summary Form, PP102A, with 2 × 4s Listed

A sample completed material summary list is shown in Figure 4-4. Assume that all the individual routine material requirements have been accumulated on the materials summary list as shown (Figure 4-4). Your next objective would be to call the local lumberyard and obtain prices. To make the task easy, the form has a column labeled "Unit Price." This column is to be used to list the individual cost of each item, which could be sold in any one of the following forms:

1. Cost per board foot.

Material summary list

Stock	Unit price	Total price	Hardware	Unit price	Total price
2 – 2 × 4 × 16'	22¢	4.69	3 lb 8d COM. NAILS	25¢	.75
3 – 2 × 4 × 12'	22¢	5.28	6 1¾" #10 WOOD SCREWS	5¢	.30
16 – 2 × 4 × 8'	22¢	18.77	1 RL 30 LB FELT	4.25	4.25
4 – 2 × 6 × 12'	26¢	12.48			
1 – 4 × 8 × ½" SHEATHING GRADE CD PLY	11.65	11.65			
10 1 × 8 × 16 No2 COMM PINE	45¢	48.60			
Total estimate		101.47			5.30

Form PP102A

Figure 4-4 Completed Material Summary List, From PP102A

2. Price per lineal or running foot.
3. Unit price, as for door sets or windows.
4. Price per sheet, as for plywood.
5. Cost per pound, as for nails.
6. Cost per roll.
7. Price of each item, as for screws and sandpaper.
8. Cost per pint or quart, as for glue and paste.
9. Cost per tube, as for chalking and glue.
10. Cost per square foot.

In every case you would list the unit cost. The unit cost is then multiplied by the quantities required per line item on the summary form. The product of the multiplication is transcribed to the total price column. (*Note:* You don't have to do the mathematics yourself. Take the completed summary form to the local lumberyard and ask the man to figure the estimate for you.)

With the material list completed and estimates obtained, the largest portion of the planning effort is complete. However, there are two significant tasks remaining: (1) listing tools that must be purchased, and (2) sequencing the tasks.

LISTING THE TOOLS REQUIRED

On the bottom of Form PP100A of the programmed plan there is a small chart. This chart is to be used to list all the tools that you *do not have* but will need to do your job. Each routine contains a list of tools. In every case the list contains the minimum tools required. These tools are generally hand tools. If you have or plan to buy portable or bench-type power tools, you can make the necessary adjustments. For instance, when using screwdrivers, a desirable substitute that makes work a lot easier is the automatic (yankee) screwdriver. This tool, shown in Figure 4-5, can be purchased with a variety of interchangeable bits.

Once all the tools you need are listed, a single call to the local hardware store will provide the cost of each. If you are estimating the cost of the entire job, expenditures for tools must be included. (*Note:* A pictorial list of hand tools that normally make up a carpentry tool kit is provided for your convenience in Appendix B.)

Figure 4-5 Yankeer Spiral Ratchet Screwdriver
(Courtesy of Stanley Tools)

TASK SEQUENCING AND TIME ALLOTMENT

A decision process must be used when compiling the task-sequence schedule. The form provided for this (Form PP102A) is found at the top of the summary sheet. It looks like the completed sample shown in Figure 4-6.

The data to be entered on this chart are obtained from page 1 of the programmed plan (Form PP100A), the section titled "Routines To Be Used." From the list of routines to be used, you must determine which routine will need to be done first, second, and third. As you arrive at a decision, transfer the routine's number and its title to the schedule. When all have been listed in the order in which you expect to do the work, refer back to the routines and multiply the estimated manhours times the number of times the routine is to be done. Indicate the total manhours (in whole hours) in the space provided on the schedule. If more than one person is required for any given routine, list the number.

If a routine is to be used more than once and at different intervals, list it as often as needed. When completed, the task-sequence schedule will provide you with a total manhour summary.

Your plan is complete. You have identified all the requirements needed for the job:

1. Title
2. Description of the job.
3. List of routines to be used.
4. List of tools you need to buy.
5. Materials needed for each routine and, where needed, the board-foot quantities.
6. Summary of materials needed, including the total estimated cost.
7. Task-sequence schedule.

Summary sheet

Task sequence schedule

No.	Title	M/H	No. men	No.	Title	M/H	No. men
BD1	EXT. DOOR UNIT INSTALL- ATION	3.5	1				
RRD2	SETTING HINGE & HANG- ING DOOR	1.0	1				
RRD3	SETTING A LOCK	.75	1				
RRD5	FITTING A THRESHOLD	1.0	1				
	TOTAL M/H	6.25					

P/O Form PP100A

Figure 4-6 Sample Task-Sequence Schedule, Form PP102A

PROGRAMMED PLAN

TITLE: _____

_____

DESCRIPTION: _____

_____

_____

_____

_____

_____

PLAN: _____

ROUTINES TO BE USED :

ROUTINE NO.	CHAP. NO.	ROUTINE TITLE	EST NO. OF MEN	NO. TIMES USED	EXT. M H

TOOL REQUIREMENT LIST

NAME OF TOOL AND SIZE	COST

FORM PP100A

MATERIAL SUMMARY LIST CONT:

STOCK	UNIT PRICE	TOTAL PRICE	HARDWARE	UNIT PRICE	TOTAL PRICE

FORM PP102A-1

SUMMARY SHEET

TASK SEQUENCE SCHEDULE

NO.	TITLE	M/H	NO. MEN	NO.	TITLE	M/H	NO. MEN

MATERIAL SUMMARY LIST

STOCK	UNIT PRICE	TOTAL PRICE	HARDWARE	UNIT PRICE	TOTAL PRICE
TOTAL ESTIMATE					

FORM PP102A

MATERIALS: ROUTINE NO. _____

QUALITY/NAME/DIMENSIONS	TOTAL REQ	BD FT

MATERIALS: ROUTINE NO. _____

QUALITY/NAME/DIMENSIONS	TOTAL REQ	BD FT

MATERIALS: ROUTINE NO. _____

QUALITY/NAME/DIMENSIONS	TOTAL REQ	BD FT

FORM PP101A-1

JOB MATERIAL REQUIREMENT LIST

MATERIALS: ROUTINE NO. _____

QUALITY NAME/DIMENSIONS	TOTAL REQ	BD FT

MATERIALS: ROUTINE NO. _____

QUALITY/NAME/DIMENSIONS	TOTAL REQ	BD FT

MATERIALS: ROUTINE NO. _____

QUALITY NAME DIMENSIONS	TOTAL REQ	BD FT

STOCK FACTOR TABLE

1x8 FACTOR 2/3 or .67	2x8 FACTOR 1 1/3 or 1.33	1x4 FACTOR 1/3 or .33	2x4 FACTOR 2 3 or .67
1x10 FACTOR 5/6 or .83	2x10 FACTOR 1 2/3 or 1.67	1x6 FACTOR 1/2 or .5	2x6 FACTOR 1 or 1.0
1x12 FACTOR 1 or 1.0	2x12 FACTOR 2 or 2.0		

FORM PP101A

5

CARPENTRY AND BUILDING TERMINOLOGY

Each field has its own language with, in many instances, words peculiar to that type of work. In addition, familiar words are sometimes used in unusual ways. For instance, the term *plate* is a building term but also indicates a container for food, a photoengraving, or a piece of steel on a ship. In building, a plate is usually a 2 x 4 (may be a 2 x 6 or 2 x 8) upon which ceiling joists rest.

To enable you to understand and interpret terms in relation to carpentry and building, this chapter discusses terms in such a way that you will recognize their application when you use any of the routines. Once you can relate certain terms to particular aspects of carpentry or of buildings, understanding their meaning will be easy.

We shall browse through a typical home with a basement, attic, and normal arrangement of rooms. The terms will be defined in depth in the appropriate usage chapter, so we will confine ourselves here to their application. Assume that we are approaching the house shown in Figure 5-1. It is a brick-veneered home with double paneled doors in the Mediterranean style. With the aid of the figure, let us define a few terms.

1. *Cornice:* along the lower edge of the roof; classified as a *boxed* or *closed cornice.*
2. *Roof:* classified as a *hip roof,* with the arrow pointing to the hip. From this view we can see four hips.
3. Roof's *valley:* an interior corner of a roof; note that there are two valleys in the view.
4. Shingles: these are *strip shingles.* Note the shingle (vertical and horizontal) alignment and the shingle application (covering) at the hip and ridge areas. The *ridge* is the uppermost horizontal line of the roof.

DOOR JAMB AND TRIM

As we enter the house we pass through the front door's *jamb, sill,* and *trim.* Figure 5-2 shows each part. Note that the jamb is located on

Figure 5-1 Front View of a Home

Figure 5-2 Door Jamb and Trim

each side and on the top, the sill is on the floor, and the trim is on both the inside and the outside. The door is hinged to the jamb, and note that a small strip of wood called a *door stop* prevents the "weather" from entering between the door and the jamb.

INTERIOR AND EXTERIOR WALL CONSTRUCTION

Stepping into the house, we encounter typical wall construction. Figure 5-3 is a cutaway view, showing the following items: (1) *shoe* (sole or sole plate) at the floor level, (2) *plate* (two each at the ceiling level), (3) *studs* in between, and (4) *drywall* or paneling fastened to the studs. On the floor level *baseboard* and *shoe molding* are installed. Where paneling is installed, a *cove* or *bed molding* finishes the ceiling-to-wall joint. On wallboard installations, *paper tape* and *bond* are used to finish the corners, and bond covers the nails.

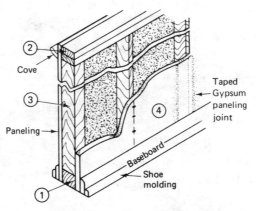

Figure 5-3 Wall Section

If we can visualize a cross section of the exterior wall for a moment we can observe (with the aid of Figure 5-4) some differences between the interior and exterior walls. The *framing* is essentially the same: *shoe, plate,* and *studs;* but whereas the inside wall is covered with *wallboard* or *paneling,* the exterior wall has full-width *insulation,* 5/8- or 3/4-in. *sheathing* and *building paper* (tarpaper). Installed on the tarpaper is brick, wood, aluminum, or vertical-panel siding or shingles.

KITCHEN AREA

Moving into the kitchen we find the usual assortment of cabinets. Notice in Figure 5-5 that the base cabinets are 36 in. high. The spacing

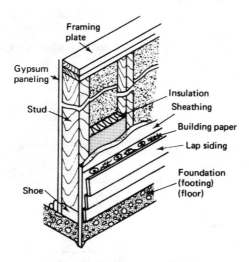

Figure 5-4 Exterior Wall

between cabinets (upper and lower) ranges from 14 to 18 in., and the upper cabinets are 30 in. high. These cabinet doors are fitted with 3/8-in. inset *self-closing hinges*. Drawers are on metal rollout runners and base cabinet shelves roll out (Figure 5-5a and b). The countertop and back splash are made from laminated plastic. Figure 5-6 identifies the cabinet facing pieces. They are the *styles* and *rails*. The styles run throughout the height and the rails run between the styles.

DOOR AND WINDOW FRAMING

All the door passages we have passed through and windows we have looked through are framed in a special way. Their basic structure is illustrated in Figure 5-7.

1. *Cripple* or *jack studs:* short pieces of studding stock used in and on door and window frames.
2. *Header:* structural member above the window.
3. *Sill* (on windows): horizontal structural member of window frame.
4. *Common studs:* full-length stud.

BASEMENT CONSTRUCTION DETAILS

Let's go to the basement. Visualize the structure of the outside wall and floor we would be standing on, as shown in Figure 5-8.

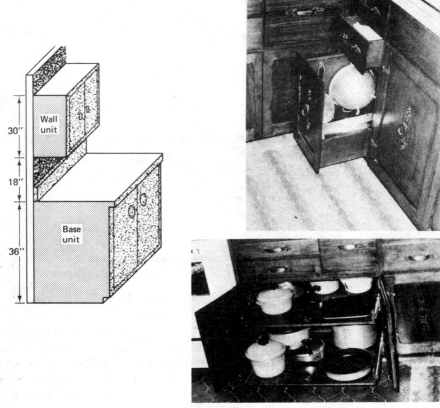

Figure 5-5 Kitchen Cabinets

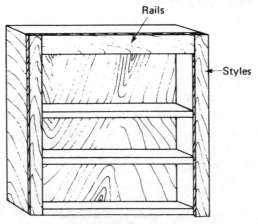

Figure 5-6 Cabinet Facing

Figure 5-7 Framing Door or Window Openings

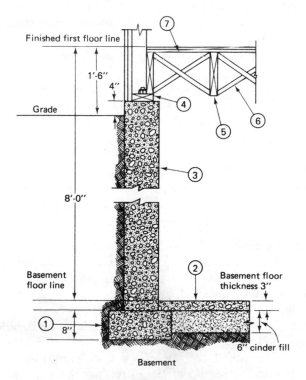

Figure 5-8 Basement

1. *Footing:* concrete reinforced with steel rods.
2. *Concrete floor:* reinforced with wire mesh.
3. *Block walls* or *poured concrete walls.*
4. *Sill:* 2 x 6 or 2 x 8 anchored with bolts.
5. *Floor joists:* for the first floor.
6. *Bridging:* for floor-joist reinforcement.
7. *Subflooring:* for first floor.

ROOF AND ATTIC CONSTRUCTION

Moving up to the attic, we find quite a few different terms. Figure 5-9 illustrates the junction of the *rafters* and *joists* to a *bearing wall.*

1. *Load-bearing wall:* wall upon which joists and/or rafters rest.
2. *Common rafter:* rafter extending from the ridge to the plate.
3. *Ridge:* horizontal member at the uppermost part of the roof.
4. *Hip rafter:* rafter reaching from the ridge (in this case) to the plate at an angle to the outside wall run.
5. *Hip jack rafter:* rafter reaching from the plate to the hip rafter.
6. *Valley rafter:* rafter reaching from the plate to the ridge, forming an inside corner.
7. *Valley cripple rafter:* rafter reaching from the valley to the hip rafters.
8. *Collar beams:* horizontal members installed to prevent or reduce spreading of bearing walls from rafters.
9. *Blown insulation:* insulation blown in place by machine.

CORNICE CONSTRUCTION

Let's get out of this cramped place and browse around outdoors. Take a good look at the cornice on the house, as viewed in Figure 5-10. As we stated earlier, it is a closed or box cornice. Its principal parts are:

1. *Frieze:* board that covers the brick (or siding) to the soffit.
2. *Soffit:* usually plywood nailed to lookouts (2 × 4 supports for soffit) and containing screen wire to ventilate the roof.
3. *Bed molding:* molding installed to complete the corner of the frieze and soffit.
4. *Fascia:* exterior board nailed over the rafter ends and extending down past the soffit.
5. *Crown molding* (or substitute): an open cornice would have the frieze board extend to the top of the rafter, allowing the rafter ends to extend to form an overhang. It would not use a soffit or fascia board.

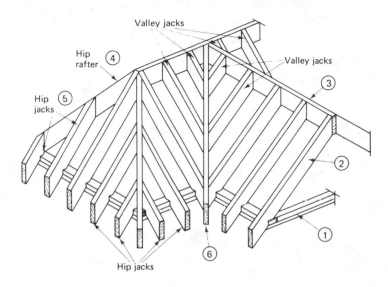

Valley jacks

Hip rafter ④

Valley jacks

Hip jacks ⑤

③

②

①

⑥

Hip jacks

Figure 5-9 Attic

Figure 5-10 Box Cornice

UNIT TWO

Basic Building

The eleven chapters of Unit Two provide detailed descriptions of the tasks necessary for building a large variety of projects. These projects range from extending a sidewalk or building a utility shed to more extensive projects, such as building a room onto a house or altering rooms; all are possible from the data provided. Each chapter contains a series of specially prepared, individually tailored rountines for your use. These 64 routines will simplify any job that you undertake. The subjects of the chapters are: cement work and foundations (Chapter 6), wall framing (Chapter 7), joist framing (Chapter 8), sheathing (Chapter 9), roof framing (Chapter 10), shingling (Chapter 11), Siding (Chapter 16), window-unit installation (Chapter 13), door-unit installation and door installation and maintenance (Chapters 14 and 15), and cornice construction (Chapter 12). All routines in this unit are prefixed with B, which indicates that the tasks all relate to basic building principles.

6

CEMENT WORK AND FOUNDATIONS

BASIC TERMS

Anchor bolts steel bolts threaded on one end and either hooked or L-shaped at the other end; embedded in concrete or between cement blocks; used to secure sills.

Batter board structure made from 2 x 4 and 1 x 6 lumber; installed at the corners of planned layouts of buildings; used to fasten foundation lines.

Concrete mixture of cement, sand, gravel, and water in specific proportions.

Floating cement act of smoothing and leveling cement or concrete with a wooden tool called a float.

Footing reinforced concrete base upon which building walls will be constructed.

Form wooden (usually) structures installed in a specified manner which will contain concrete in liquid form until it hardens.

Plyform exterior-type plywood specially prepared plywood of excellent quality, made with exterior glues, designed to be used for form work; may be reused often if properly conditioned.

Screeding act of leveling concrete with the aid of stock lumber and a push-pull and forward motion.

Steel reinforcing rods steel rods sold in varying diameters and lengths which can be fairly easily bent; used to reinforce footings and other areas of cement that are expected to carry heavy loads.

Trowel rectangular flat metal tool with a handle in the center used in concrete work (Figure 6-1).

Wire mesh heavy-gage wire usually meshed at 6 in. x 6 in. intervals; used to reinforce sidewalks, patio floors, room floors, and driveways.

This chapter presents details and routines for a variety of tasks that involve the use of concrete. These include the installation of sidewalks and curbs, the laying of steel, and the pouring of concrete slabs. The descriptions of the routines are sufficient to do professional work. However, if a room is being added to your home, for example, it may be

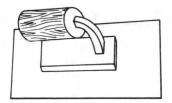

Figure 6-1 Metal Trowel

advisable to contract for the work, as the exactness required may be above your capability and allotment of time. The finished floor must be level, even, and finely troweled. On the floor for an average-sized room, this is not easy to do. However, let's drop the negatives and proceed with what we can do.

SCOPE OF THE WORK

A considerable amount of time, work, and material will be used in the preparatory stages of cement work. Basic areas such as footings, slab areas, sidewalks, and curb trenchings must be laid out. Frequently, the layout work involves erection of batter boards and foundation lines. Following these efforts, forms of varying sizes and shapes must be installed to contain the concrete until it sets (hardens). The steel rods and wire mesh used to strengthen concrete are placed into the formed area prior to pouring the concrete. Mixing and pouring concrete is strenuous work, but the procedure is relatively simple. When pouring, the concrete must first be evened with a board, then floated to make it level, and finally troweled smooth or broomed rough, as required. After a drying period of 3 to 7 days, the forms may be removed.

Concrete is made from cement, sand, gravel, and water. The mixture consists of a specific combination of all these elements. The various combinations of cement, sand, and gravel are:

Type A: 1 part cement
2 parts sand
2 parts gravel

Type B: 1 part cement
2 parts sand
4 parts gravel

Type C: 1 part cement
3 parts sand
5 parts gravel

Each type has specific qualities: the strongest is type A; type C has the least strength but the greatest coverage.

The dry products are added proportionally in a box, wheelbarrow, or on to a large flat area, such as a sheet of plywood. A hoe is the best tool to use to thoroughly mix the dry materials. After a uniform gray color is obtained, water is added in controlled quantities until the mixture is wet but not runny. At this time the concrete is ready to pour into a previously prepared form.

The first group of tasks that you will use is that required to install the batter boards along foundation lines, for footings, and for floor forms. The second group of tasks is that required to install steel rods and wire mesh and for the pouring and finishing of cement.

INSTALLING BATTER BOARDS

Each corner of a planned building or addition to a home will require a batter-board assembly. Cut three pieces of 2 x 4 36 to 48 in. long for each corner, then taper one end of each to a point (like a pencil). Next cut two pieces of 1 x 6 36 to 48 in. long for each corner.

As illustrated in Figure 6-2, measure off, from a property line, existing house line, or center of the street, two reference points that will establish the foundation line. Drive two 2 in. x 2 in. x 12 in. stakes into the ground to mark the reference points. Next, determine the relative position of the new work along the reference line. Lay out and drive a 2 x 4 stake into the ground approximately 18 in. on either side of the reference line and 18 in. back from the proposed outer end of the foundation (refer to Figure 6-2A). Drive the third stake at approximately a right angle and 36 to 48 in. from the corner 2 x 4. Measure the approximate length of the foundation and install another batter-board assembly.

Locate and transfer a height reference point from an existing foundation, street height, or ground height by using a mason line and a line level (Figure 6-3). Use a transit if one is available. After fastening the line against the reference point, and with the level installed at the midpoint of the line's length, have a helper read the level while you hold the other end of the line against a batter-board stake. Adjust the line up or down against the stake and mark the level point. Install the top of a 1 x 6 even with the mark on the 2 x 4. After leveling the 1 x 6 with a 24-in. level, nail it to the second 2 x 4. Install the other board to complete the batter-board assembly. Repeat this process for the second corner. Next, install a line over the ground-staked reference points by partially driving a nail into the top of the 1 x 6 at a point perpendicular to the ground stake. Use a plumb bob on a line to obtain this point.

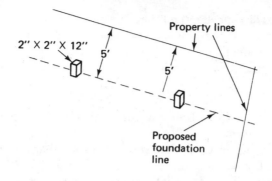

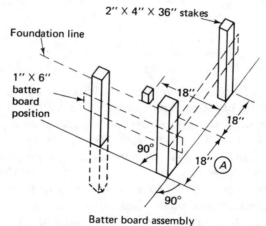

Figure 6-2 Laying Out a Foundation

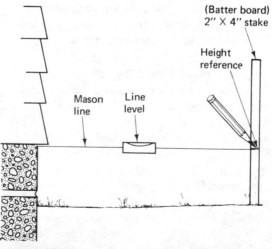

Figure 6-3 Locating a Height Reference

INSTALLING FOUNDATION LINES

When the reference line has been installed, determine the line 90 degrees from it at each end of the proposed foundation. Use the 3-4-5 method (Figure 6-4). Install one end of a line onto the adjacent batter board and adjust its movement right or left until the 3-4-5 figures are achieved. Mark this position with another 2 in. x 2 in. x 12 in. stake at the length of the side wall. Install batter boards as before. Repeat the process for the fourth corner. Install the four lines.

Now you must check that your lines are square. Use a tape measure to measure diagonally across the area to the intersect points of the lines. If the distances are equal, the foundation lines are properly installed; if not, you have a parallelogram. Use the 3-4-5 method again and square one corner; readjust your lines accordingly until you are successful.

The next phase is to dig out for the footings.

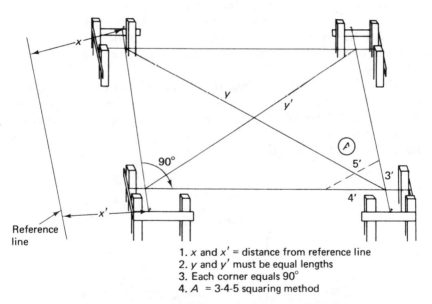

1. x and x' = distance from reference line
2. y and y' must be equal lengths
3. Each corner equals 90°
4. A = 3-4-5 squaring method

Figure 6-4 Foundation Lines and the 3-4-5 Method

INSTALLING FOOTING FORMS

Sometimes it is possible to use the earth as boundaries for footings. By carefully digging the two sides and the bottom, the finished footing area is in firmly packed earth, as shown in Figure 6-5. If the earth will not support the concrete, wood forms must be installed, as detailed in

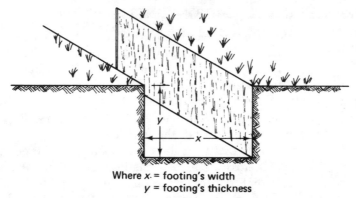

Where x = footing's width
y = footing's thickness

Figure 6-5 Footing, Using Firm Ground as Forms

Routine BC3 at the end of the chapter. Boards must be installed parallel, staked, and nailed in precise position. An average footing is 16 in. wide and 8 in. deep and looks as shown in Figure 6-6. The outside form may be in line with the finished line or 4 in. or more outside the finished line. The height of the board is measured from the lines installed on the batter boards.

After all the outside footing boards along one wall are installed, the inside forms are installed by using spacer sticks and a level. When completed, earth is backfilled against the forms. This earth aids in making the forms more stable. Repeat these steps to prepare footings for other walls or sides of the work that need them.

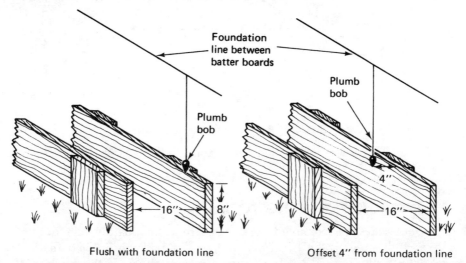

Flush with foundation line   Offset 4″ from foundation line

Figure 6-6 Types of Footings

Corners are made by cutting the inside form shorter than the outside form by the width of the footing. The form material should be nailed at the corners, and a stake should be driven on each side of the corner.

INSTALLING FLOOR, SIDEWALK, AND DRIVEWAY FORMS

Floor and sidewalk thicknesses range from 3-1/2 to 6 in. On the average a slab patio or floor will be 3-1/2 to 4 in. thick. A driveway is usually 6 to 8 in. thick, to safely bear the weight of cars and trucks. Forms for these jobs are usually made from 2-in. stock lumber. The forms are installed similar to the footing boards (Figure 6-7). The stakes that hold the forms erect should be at least 15 in. long. A rule of thumb for this length is: three times the depth of the concrete slab in medium- to hard-packed earth; four to five times the depth of the concrete slab in loose, sandy soil.

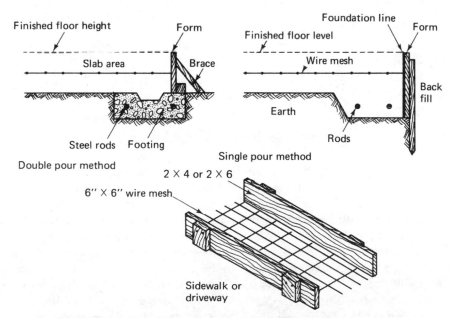

Figure 6-7 Forms for Floors, Sidewalks, and Driveways

As a rule, the height of the form will be determined from the batter-board lines. In some installations, such as patios, slopes to the floor are needed. In these jobs install the forms against a reference height and slope the form away from the reference point. Usually a 1- to 1-1/2-in. slope in 12 ft is adequate for proper drainage.

With the forms installed, the next phase of the program can begin.

INSTALLING STEEL RODS AND WIRE MESH

Both steel rods and wire mesh are used in concrete work. Each has its own special application in projects performed within the scope of this book. The rods are used in footings and the wire mesh is installed in flat slab areas. The rods are suspended with bailing wire, nailed to the form (Figure 6-8), and should be installed at a distance from the bottom equal to one-third to one-half the thickness of the form.

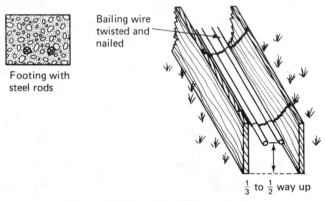

Footing with steel rods

Bailing wire twisted and nailed

$\frac{1}{3}$ to $\frac{1}{2}$ way up

Figure 6-8 Steel Rods in a Footing

For a good job, the rods are overlapped at the corners; for a better job, the rods are bent at the corners and joined along the straight runs. Figure 6-9 shows both methods. Note in the cross lap that all four intersects are tied with bailing wire at the corners. In the other application, the rod ends are overlapped 4 to 5 in. and tied with bailing wire. To make this installation possible, the rod must be bent at a 90-degree angle. A simple jig for bending the steel can be quickly and easily made from available stock. Use Figure 6-10 as an aid in making such a jig. Nail a 2 x 6 across two sawhorses. Nail two scrap pieces of 2 x 4 on the 2 x 6 as shown on the figure. Allow a space between the 2 x 4s equal to the diameter of the steel. Insert the rod to a point where the bend will be needed, then walk the free end back by pulling the rod until the 90-degree bend is made. Concrete slabs are usually reinforced with 6 x 6 in. steel wire mesh. This mesh is sold by the running foot in standard 6-ft widths. Two people are usually needed to handle the wire mesh; it is heavy and

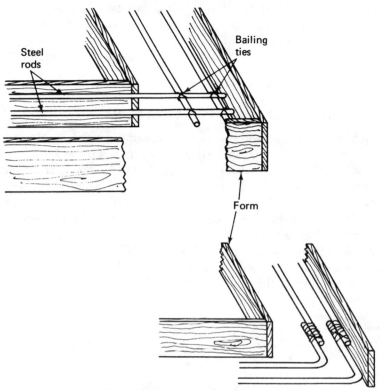

Figure 6-9 Tying Steel Rods

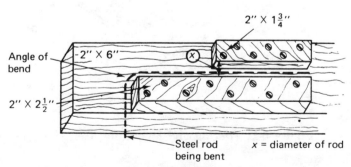

Figure 6-10 Steel-Rod-Bending Jig

can cause injuries. When previously rolled, then unrolled, it can reroll itself and hurt an unwary workman.

A pair of bolt cutters is ideal to cut mesh but it will be more economical to rent a pair than to buy them unless a significant amount of concrete work is to be done. Wire mesh is flattened, cut, and laid in the confines of the form. An overlapping technique should be used and the overlapped pieces secured with bailing wire.

Once the steel rods and wire mesh are installed, the concrete can be ordered.

POURING AND SMOOTHING CEMENT

A hoe or rake and a shovel are needed to move liquid concrete within a form. While the concrete is being poured into the form, it should be leveled with a rake or shovel (Figure 6-11). A rake is best. After all the concrete is poured and leveled with the hoe or rake, screed it by holding the screeding board with both hands and its lower edge on the opposing concrete forms. Slide the screeding board back and forth while dragging it and the excess concrete toward you. A person on each end of the screeding board makes the work easier. Follow the screeding with a floating process, using a wooden float. This levels out the surface, buries the gravel, and brings the cement to the surface.

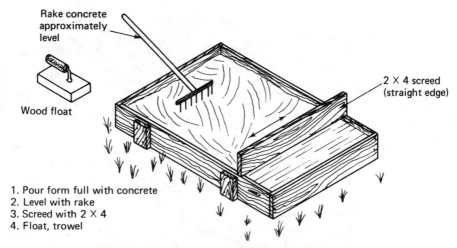

Figure 6-11 Pouring and Smoothing Cement

POLYURETHANE UNDERLAYMENT

All areas that will have heating, air conditioning, resilient tile, or rugs should have an underlayment installed directly on the ground within the form. This product is sold in 4-, 6-, and 8-ft widths and in 4- and 6-mil thicknesses. The 6-mil thickness is adequate for most jobs. Overlap joints by 18 in. and allow the ends and sides to extend over the forms.

INSTALLING ANCHOR BOLTS

An anchor bolt is a specially designed bolt inserted into concrete to which a wall area will ultimately be secured. The bolts are available in various lengths and usually are shaped as a J or an L.

After the slab (or footings in some cases) is poured and has begun to set, the bolts must be worked into the concrete (Figure 6-12). The bolts are usually installed 2 in. in from the edge of the slab and to a depth of all but 2-1/2 in. This 2-1/2 in. allows for a 2 in. x 4 in. sill and 1 in. of thread for the washer and nut.

The bolts are spaced along the walls approximately 15 in. from the corners and 12 in. from each side of door openings. In long spans they are spaced at 6- to 8-ft intervals.

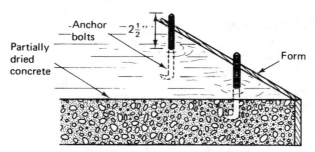

Figure 6-12 Installing Anchor Bolts

ROUTINES

Next, 10 routines are given which provide the instructions needed to do all the jobs described and similar tasks. When you have work of this kind to do, select the routines that will fulfill your requirements and enter them into the programmed plan in Chapter 4.

BC1: INSTALLING BATTER BOARDS

RESOURCES

Materials:
3 2 x 4s 3 to 4 ft long per corner

2 1 x 6s 4 ft long per corner
1/2 lb 6d or 8d common nails per corner
Floor plans: foundation layout and elevation reference drawing

Tools:
1 no. 8 crosscut handsaw
1 hand ax
1 10- to 16-lb sledge hammer
1 24-in. level
1 16-oz claw hammer
1 line level (not needed if a transit is used)
1 100-ft mason line
1 100-ft tape (50-ft tapes may be used on small jobs)

ESTIMATED MANHOURS

For two people   4 to 8 hours

PROCEDURE

Step 1   Prepare 2 x 4 stakes by cutting to a specified length and tapering one end to a point with the hand ax.

Step 2   Precut 1 x 6 stock to 48-in. lengths.

Step 3   Measure back from the foundation line (use your plan to obtain the foundation line's position).

Step 4   Drive one stake approximately 18 in. either side of the line about 18 in. back from the corner to be made.

Step 5   Drive the third stake at a point 90 degrees from the first two stakes and 36 in. away from the corner stake, forming a right angle.
Repeat steps 3 through 5 at an adjacent corner of the proposed foundation.

Step 6   Using the mason line and line level, fasten one end of the line to the reference height you will use (see your plan). Connect the line level to the mason line at the midpoint of the span. While one person reads the level, the other person adjusts the line held against one of the 2 x 4s (previously driven into the ground) until the bubble is centered. The position is then marked on the 2 x 4.

Step 7   Nail the 1 x 6 x 48 in. pieces to the 2 x 4s with the tops of the 1 x 6 even and level with the mark made in step 6.

Step 8 Repeat steps 6 and 7 for the second corner, and repeat the process for all remaining corners.

BC2: INSTALLING FOUNDATION LINES ON BATTER BOARDS

RESOURCES

Materials:
Floor plans
100 ft or more of mason line
2 lb 8d common nails

Tools:
1 100-ft or 50-ft tape
1 16-oz claw hammer

ESTIMATED MANHOURS

2-1/2 hours

PROCEDURE

Step 1 Measure from the common reference point to the two batter boards (this may be a property line, a house line, or the center of the street).

Step 2 Connect a mason line between the two points after partially driving nails into the marks made on the batter boards.

Step 3 Measure the wall length from the first mark to the next batter board. Drive a nail into the mark. (Use your plan to obtain this length.)

Step 4 Repeat step 3 at the other end of the layout. Connect a line between the marks (see Figure 6-4).

Step 5 Measure from another reference point for the walls, which will run perpendicular to the ones just defined. Mark each point and drive a nail into each mark. (Again use your plan.)

Step 6 Connect the nails with a line (or lines).

Step 7 Check for square at the corners.

a. Make a mark on the line 3 ft from the corner and 4 ft from the same corner on the adjacent line.

b. Measure the distance (hypotenuse) between the 3- and 4-ft marks. If it is 5 ft, your corner is square.

Step 8 Check for square diagonally.

a. Measure diagonally across the corners.

b. If measurements are equal, foundation lines are properly installed; if not, adjust accordingly.

BC3: CONSTRUCTING FORMS FOR FOOTINGS

RESOURCES

Materials:

Floor plan and specifications, especially footing detail

18 to 24 in. of 1 x 3 for _____ stakes at 4-ft intervals, plus 1 stake
                             no.

for each inside and outside corner. The number of stakes × the length (18 or 24 in.) divided by 12 equals the number of lineal

_____ ft of 1 in. material × _____ footing material, plus 1
lineal   width

stake for each inside and outside corner. (*Note:* 2-in. material makes a better form, but 1-in. material is cheaper and works well if backed by earth. The average width of form stock is 8 in.)

3 to 5 lb of 6d common nails per 50 ft of footing

Tools:

1 24-in. level
1 2-lb sledgehammer
1 16-oz claw hammer
1 mason line (average 100 ft)
1 no. 8 crosscut handsaw
1 pair sawhorses

1 6-ft folding ruler
1 flat shovel
1 plumb bob with line

ESTIMATED MANHOURS

For 16 lineal ft   3.5 hours
plus corner   1.0 hour
(does not include digging)

PROCEDURE

Step 1 Precut the first of two structural members to the length of the *outside* footing if *less* than 16 ft long.

Step 2 Dig and level the earth below the foundation line as required by the specifications.

Step 3 Lay the footing board in the cleaned area.

Step 4 Tie a plumb bob to the foundation line and from its point measure outward a distance shown as A in Figure BC3. Butt the footing board against the end of the ruler.

Alternative step 4 Where the footing is flush with the outer wall, install the inside of the form even with the point of the plumb bob.

Step 5 While holding the board erect, position a stake along the outside surface of the footing board and drive it in with the sledgehammer.

Step 6 Repeat steps 4 and 5 at the other end of the footing board.

Step 7 Take two small 1 in. x 2 in. x 12 in. stakes and drive them directly below the foundation line to the height of the top of the footing at each end of the footing board.

Step 8 Position the level on top of the stakes just driven in and the footing board. Nail the board to its stake after adjusting the board for a level position.

Step 9 Repeat step 8 at the other end of the footing board.

Step 10 Sight the installed footing board and, when straightened, drive additional 1 x 3 stakes at 4-ft intervals. Nail the stakes to the board.

Step 11 Repeat steps 3 through 10 until all outside footing boards are installed.

Step 12 Cut two or three pieces of 1 x 3 or 2 x 4 equal to the width of the footing (e.g., for a 16-in. footing width, cut 16-in. pieces).

Step 13   Precut the footing board, if necessary.

Step 14   Lay the board in the footing and place a 1 x 3 or 2 x 4 spacer between the footing boards. Drive a stake on the outside of the board being installed.

Step 15   Repeat step 14 at the other end of the board.

Step 16   Level across the two opposing footing boards and, when level, nail the board to the stake. Repeat the process at the other end of the board.

Step 17   Repeat steps 13 through 16, taking special care to form the corner.

Step 18   Backfill with earth on the outside of the forms and pack slightly.

Step 19   Remove the guide stakes that were driven between the footings.

BC4: CONSTRUCTING FORMS FOR FLOORS

RESOURCES

Materials:
Floor plans: cement, floor details

_____ lineal ft of 1-in. material × _____ for forms. (*Note:*
no.   width

2-in. stock may be used.) The *width* of the stock is equal to the thickness of the concrete floor *and* the footing requirement. The number of lineal feet equals the length of the perimeter of the floor.

_____ ft of 1 x 3 for stakes. The average should be one 18 to 24-in.
lineal

stake per 4 ft of form plus 1 extra for each corner.

_____ 6d common nails (average 1 lb per 20 lineal ft of form)
lb

Tools:
1 24-in. level
1 16-oz claw hammer

1 2-lb sledgehammer
1 100-ft mason line
1 pair sawhorses
1 no. 8 crosscut handsaw
1 plumb bob

ESTIMATED MANHOURS

For 20 lineal ft   3.0 hours

PROCEDURE

Preliminary step 1   Prepare the ground by leveling and trenching around the perimeter.

Preliminary step 2   Install batter boards (Routine BC1).

Step 1   Precut the form stock material to a length needed for one side.

Step 2   Precut the stakes and taper to a point.

Step 3   Lay the form piece in the trench.

Step 4   Tie a plumb bob to the foundation line (previously strung).

Step 5   Align the inside of the form with the tip of the plumb bob. Place a stake flush with the outside of the form and drive into the ground with the sledgehammer.

Step 6   Repeat steps 4 and 5 at the other end of the form.

Step 7   Nail the form to one stake *after* positioning the form for the correct height measured from the batter-board line.

Step 8   Place the level on the top edge of the form at about its mid-length. Adjust the free end of the form for level. Nail to the stake.

Step 9   Install additional stakes at 4-ft intervals and nail to the form.

Step 10   Repeat steps 1 through 9 for each remaining side of the form.

Step 11   Tie the corners together by nailing through the forms.

Step 12   Backfill with earth. Loosely pack the earth; *do not* disturb the form.

Test: Check for square using the 3-4-5 method and the diagonal method.

BC5: CONSTRUCTING FORMS FOR SIDEWALKS AND DRIVEWAYS

RESOURCES

Materials:

_____ lineal ft 1 x 4 or 2 x 4 for form
   no.

_____ lineal ft 1 x 3 for stakes. Each stake equals 15 to 18 in. and
   no.

one is required for each 4 lineal ft.

One piece of 1/2 in. × 3 in. expansion material (fiberboard) for each
8 to 12 ft

Tools:
1 100-ft mason line
1 6-ft folding ruler
1 13-oz claw hammer
1 24-in. level
1 no. 8 crosscut handsaw
1 hand ax

ESTIMATED MANHOURS

For 12 lineal ft   1.5 hours
(does not include digging and leveling)

PROCEDURE

Preliminary step Prepare the earth for the sidewalk by leveling and/or sloping.

Step 1 Drive two stakes (one at each end) along a line where one edge of the sidewalk will be installed.

Step 2 String a mason line between the stakes 6 in. higher than the finished height of the sidewalk.

Step 3 Position a 1 x 4 or 2 x 4 under the line; align the inner edge of the stock with the mason-line position and drive a stake *behind* the stock.

Step 4 Repeat step 3 at the other end of the stock.

Step 5 Measure with a ruler 6 in. between the top of the form and string that is nailed to the stakes.

Step 6 Repeat steps 3 through 5 until the full length of the sidewalk form is installed.

Step 7 Cut two spacer sticks (use 1 x 4) as long as the width of the sidewalk.

Step 8 Place the sticks perpendicular to the first form installed and *butted* to it.

Step 9 Lay the opposite form member against the other end of the spacer stick. Drive a stake on the outside of the form material.

Step 10 Repeat step 9 at the other end of the form piece.

Step 11 Place a straightedge across the opposing form and level on top of the straightedge.

Step 12 Adjust the form beneath the straightedge until the level's bubble indicates level. Nail the stake to the form.

Step 13 Repeat steps 8 through 12 until all the forms are installed.

Step 14 Install pieces of 2 x 4 across the ends of the forms. Backfill with earth.

Step 15 Insert the expansion material, cut X the width of the sidewalk or driveway, each 8 to 12 ft

BC6: CONSTRUCTING FORMS FOR CURBS

RESOURCES

Materials:

_____ lineal ft of 2 x 4 stock equal to the length of the curb
   no.

_____ lineal ft of 1 x 6 stock equal to the length of the curb
   no.

_____ sheets B-B plyform exterior-type plywood. One sheet 4 ft x 8
   no.

ft is enough for one side of 32 lineal ft of curb.

———————— lineal ft of 1 x 3 stock for stakes. Each stake is 23 in. long
 no.

and the stakes are spaced 4 ft apart.

1 lb 8d common nails per 8 ft of form

1 lb 6d common nails per 8 ft of form for nailing stakes

1 lb 2-1/4-in. (8d) hardened cut nails for nailing 2 x 4 to driveway surface

Tools:

1 no. 8 crosscut handsaw

1 16-oz claw hammer

1 24-in. level

1 mason line

1 2-lb sledgehammer

ESTIMATED MANHOURS

For 16 lineal ft   2.0 hours
(does not include digging)

PROCEDURE

Preliminary step   Dig a trench one shovel wide and approximately 6 in. deep.

Step 1   Prepare the **L**-shaped pieces needed (see A, Figure BC6) by nailing the 1 x 6 to the 2 x 4 with no. 8 common nails. Keep the bottom edge of the 1 x 6 and the flat bottom side of the 2 x 4 flush. Cut for length as required.

Step 2   Nail the **L** form along the edge of the driveway (or other cement/asphalt surface). Repeat steps 1 and 2 as often as needed until the task is complete.

Step 3   Cut the sheet of plyform into four 11-7/8-in. × 96-in. strips.

Step 4   Cut two or more spacer sticks equal to the width of the finished curb.

Step 5   Position each piece of 11-7/8-in. × 96-in. plyform opposite the **L** strip. Insert the spacers and drive the stakes into the ground on the *outside* of the plywood form (one per 4 ft).

Step 6   Adjust the plywood strip for level with the **L** strip by using a level across the two pieces. Nail the stake to the plywood with 6d common nails.

Step 7 Backfill with earth.

BC7: INSTALLING STEEL

RESOURCES

Materials:
_____ lineal ft of 3/8- or 1/2-in. steel rod (select one size)
　　no.

50-ft roll 16- to 18-gauge bailing wire
2 lb 4d common nails

Tools:
1 pair 6-in. lineman's pliers
1 13-oz claw hammer
1 hacksaw
1 bending jig
1 piece white chalk
1 pair bolt cutters (optional)

ESTIMATED MANHOURS

16 lineal ft (2 rods)　　**2 hours**

PROCEDURE

Preliminary step Manufacture a jig (see Figure 6-9).

Step 1 Lay out the steel rod in the form. Mark for bend with white chalk, and bend in jig.

Step 2 Lay out, mark, and bend the second steel rod (if used).

Step 3 Suspend the steel rod in the form by laying it on top of some scrap material (e.g., brick or a short piece of 2 x 4).

Step 4 Cut two pieces of bailing wire sufficiently long to reach across the form.

Step 5 Nail one end of the pair to the top of the form (bend the nail over rather than driving it home) and begin to twist the pair of wires.

Step 6 Make twists before, between, and after each piece of steel rod.

Step 7 Nail the other end of the pair of wires to the top of the form.

Step 8 Repeat steps 1 through 7 until all the steel is installed, and steps 4 through 7 each 4 to 5 ft.

Step 9 Cut a 12-in. piece of bailing wire for each overlap of rod ends. Use the lineman's pliers to tie the ends together. Trim off the excess wire.

Step 10 Remove the scrap material holding the steel aloft.

BC8: INSTALLING WIRE MESH

RESOURCES

Materials:
_____ sq ft of 6 in. x 6 in. wire mesh. The number of square ft equals
        no.

the size of the concrete area to be poured, *plus* 2 percent for overlap and end waste.
25 ft of 16- or 18-gauge bailing wire

Tools:
1 pair bolt cutters
1 pair lineman's pliers (if tying is to be done)

ESTIMATED MANHOURS

For 100 sq ft   1.5 hours
(an assistant is desirable)

PROCEDURE

Step 1 Unroll the wire mesh and bend it to make it lay flat.

Step 2 Measure and cut each piece needed.

Step 3 Lay each precut piece in place.

Step 4 Tie all overlap areas with bailing wire. Trim the excess bailing wire.

BC9: POURING AND SMOOTHING CONCRETE

RESOURCES

Materials:

_____ cu yd of concrete mix. One cubic yard covers 27 cu ft, or
   no.

an area:
    27 sq ft 12 in. thick
    27 lineal ft 8 in. thick x 16 in. wide
    54 sq ft 6 in. thick
    81 sq ft 4 in. thick
   108 sq ft 3 in. thick

Tools:
1 2 x 4: 8, 10, 12, or 16 ft (for screeding)
1 wooden trowel
1 steel trowel
1 garden rake
1 square-point shovel

ESTIMATED MANHOURS

3 hours per cubic yard

PROCEDURE

Step 1 Have the concrete poured into the center of the form, away from form board.

Alternative step 1 Mix the concrete as described in the chapter.

Step 2 Rake the concrete toward and against the forms. Fill all the crevices.

Step 3 Tamp the concrete into and around the forms.

Step 4 Use a 2 x 4 to screed the concrete. Use a back-and-forth stroke while constantly moving forward. Repeat as often as necessary to level the concrete.

Step 5 Float the surface of the cement with a wooden float. (Repeat this step again after the concrete has dried somewhat.)

Step 6 For a smooth, polished finish, trowel the surface with a metal trowel, adding water in a sprinkling fashion if moisture is needed.

Step 7 Broom the surface with a push broom to roughen the surface for driveways and sidewalks.

Step 8 Wait 3 days or longer, then remove the forms.

BC10: INSTALLING ANCHOR BOLTS

RESOURCES

Materials:
_____ of anchor bolts. The bolt's length is 8, 10, or 12 in. or longer,
   no.

bent at the bottom end.
Floor plans

Tools:
1 trowel or float or both
1 ruler or 50-ft tape

ESTIMATED MANHOURS

5 minutes per bolt
plus 15 minutes per side of form for layout

PROCEDURE

Step 1 Lay out the placement of bolts.

a. Use a floor plan and a tape or ruler.

b. Mark a form for each door opening.

c. Mark in from each corner and on each side of the door opening 12 to 15 in.

Step 2 Push a bolt into the concrete at the mark and 2 in. in from the form.

Step 3 Hold the end of the bolt 2-1/2 in. above the concrete. With the trowel, smooth the concrete around the bolt.

Step 4 Repeat steps 1 through 3 until all the bolts are installed.

Caution: Recheck the height of the bolts after 30 minutes, as they may slide into the cement.

WALL FRAMING

BASIC TERMS

Stud normally a 2 x 4 used as part of a wall, reaching from floor to ceiling; a vertical member of a wall.

Shoe base of a wall unit, sometimes called a *base plate* or *sole;* the part of the wall unit to which the bottoms of the studs are nailed.

Plate top horizontal member of a wall unit; the part of the wall unit to which the tops of the studs are nailed.

Jack stud piece of vertical wall member nailed to either shoe or plate but not both, or between the header and sill on the window frame.

Header structural member usually made from two pieces of stock, spanning door or window openings.

Toe nailing method of nailing whereby the nail is driven at an angle through the end of one board into the second member.

Framing a wall with 2 x 4s and other structural members is an enjoyable and relatively simple job. It is a job that frequently requires the aid of a helper, especially during the raising of the wall sections.

SCOPE OF THE WORK

The eight routines at the end of the chapter provide a clear, easy way to prepare and erect walls. Although the routines provide a logical sequence, it is not necessary in all cases to use them in the sequence listed. Selected routines should be used when remodeling and repairing. When remodeling, a need often arises for constructing a new wall or making a door opening. In repair tasks, including framing repairs, the data contained in these routines will provide the information required to effect the repair. Keep these ideas in mind as we examine the principles of framing walls.

WALL LAYOUT AND CONSTRUCTION

Open your plans to the floor plan view or to the sample floor plan presented in Chapter 2. Consider each wall as a single item. With this approach you can plan your work. Because of tradition and because it makes the work easier, the outside walls are laid out, built, and erected first. Following this task the inside room walls are laid out, built, and erected. Finally, the closet walls are installed.

The shoe for a single wall may be laid in place and secured, or all the shoes may be laid in place and secured within the same time period. The plate for each wall should be cut and laid alongside its mating shoe. Where walls are longer than 16 to 20 ft, the 2 x 4s must be spliced. In these cases the plate 2 x 4s must be cut so that the splice falls in the center of a stud (Figure 7-1).

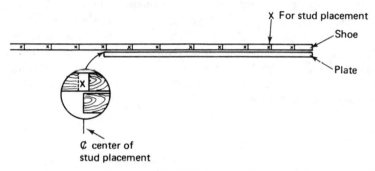

Figure 7-1 Splicing a Plate

Lay out all door and window openings according to your plan. Where possible, always work from the same end of the wall. Using your framing square or a 50-ft tape, lay out all common-stud positions for either 16 or 24 in. OC (OC, on center) as the plan dictates. Assuming a wall with studs spaced 16 in. OC, measure from the end of the shoe *in* a distance of 15-1/4 in. (Figure 7-2). Draw a line across both the plate and the shoe. Then place an X on the far side of the line. The X will be the position of the stud. The line will be your aid when you finally nail the stud in place.

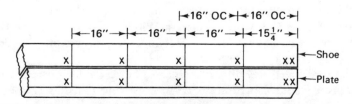

Figure 7-2 Laying Out Studs 16 in. On Center

Why should the first measurement be only 15-1/4 in. from the end? Remember, we want the studs spaced 16 in. OC; therefore, we must take into account the thickness of a stud. Since a stud is 1-1/2 in. thick, one-half of its thickness equals 3/4 in. All this results in efficient use of the sheathing on the outside of the wall. It also results in efficient use of gypsum wallboard and paneling where inside walls are installed.

If your plan calls for studs spaced 24 in. OC, your first measurement will be 23-1/4 in. and all other measurements will be 24 in. apart.

Routine BF1 outlines the materials, tools, estimated manhours, and procedures that you will use in constructing your walls. But, let's continue with the task.

Use your plan to determine the size of the window and door openings in the section of wall on which you are working. Measure this distance on the shoe and plate. For example, look at Figure 7-3. A 36-in. window is to be installed. Because some leeway or space is needed to account for swelling of stocks, about 1 in. of extra space is provided in the opening. Therefore, as shown in Figure 7-3, the layout is for 37 in., not 36 in. (18-1/2 in. each side of the reference center line).

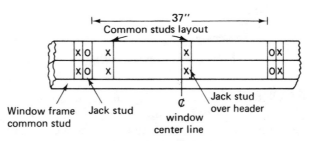

Figure 7-3 Window or Door Layout

With your wall layout complete, cut a stud of common length for each place marked on the shoe. This stud length will be found on your plan. As a rule, ceilings are set at a height of 8 ft 1 in. Therefore, the stud will be 8 ft 1 in. less shoe, and two plates or 4-1/2 in, or 8 ft 1 in. = 97 in. less 4-1/2 in., which equals 92-1/2 in. long.

Before assembling a section of wall that contains a window or door, you might want to fabricate the frame for them. In this case you would use either Routine BF3, Constructing a Window-Frame Unit, or Routine BF4, Constructing a Door-Wall Unit. If there are no windows or doors in the wall, proceed to nail each stud to the plate.

Before raising the wall, construct the outer corner posts if they are required. You would use Routine BF2, Preparing Outside Corner Posts. Corner posts are easily made. They usually require nailing three 2 x 4s together plus the addition of spacer 2 x 4 pieces. When these 2 x 4s are properly assembled, they not only create an outside corner but make an inside corner as well. Figure 7-4 shows how the corner looks from the end view. When completed, position the corner post or posts in the wall unit and nail to the plate.

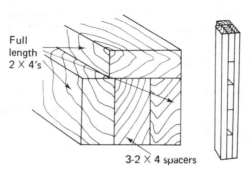

Full length 2 X 4's

3-2 X 4 spacers

Figure 7-4 End View of Outside Corner Post

Recall that in the example a window opening for a 36-in. window was laid out. That, of course, means that Routine BF3 would be used, because it outlines construction of a window opening. The easiest way to perform this task is by cutting the pieces, assembling them on the deck, and then installing the assembly in place within the wall before it is raised.

So what is involved? First, a header must be constructed. In our example it must be: 37 in. for the finished opening plus 3 in. for the jack studs (1-1/2 in. thick × 2 studs), for a total of 40 in. Since the 2 x 6s (header stock) are only 1-1/2 in. thick and a 2 x 4 is 3-1/2 in. thick, a spacer of 1/2 in. stock cut from 2 x 4s or 1/2 in. plywood must be used to fill out the header. Spacers for 2 x 6 headers are usually 1/2 in. x 1-1/2 in. x 5-1/2 in. They are nailed into place so that when the header is complete, its total depth equals 3-1/2 in.

The spacers play an important role. In order for sheathing to be nailed along an even surface on the outside of the wall, the header surface must be nailed even with the outer edges of the studs. Without spacers the header would have a 3-in. depth. Nailing wallboard or paneling above the window or door on the inside of the wall would be impossible. More significantly, installation of window or door trim would not be possible because of the 1/2-in. gap between the wallboard and the 2 x 6 header.

With the header completed, a sill made from a single 2 x 4 (or doubled, if desired) is cut. Its length is the same as the length of the header.

Next the position of the header and sill are marked on two common studs. The positions of these two elements must be determined from the *elevation* plan and marked on the common studs. The window opening must be the height of the window plus 1 in. (as a rule). After marking the head and sill position, square these marks across the face of the 2 x 4s.

Position the header and sill between the two common studs where each belongs and nail through the common stud into the header and sill. While the frame is on the deck is an ideal time to cut and install the jack studs. Measure the distance from the header to the end of the common stud. Cut and nail all needed jack studs. One is nailed to each common stud and one is toe-nailed where each common stud would *normally* be located. Repeat the process stated above for all jack studs below the sill. Finally, cut and nail two jack studs between the sill and header. Your completed frame will look as shown in Figure 7-5.

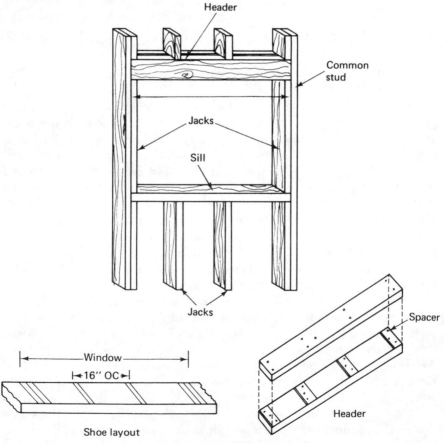

Figure 7-5 Window-Frame Unit

Lift the window unit and place it within the wall unit. Nail all studs to the plate with 12d or 16d common nails. Be sure to align the studs on the proper marks.

RAISING THE WALL

You are now ready for help, because the wall is ready to erect. A common occurrence when raising walls is that everyone lets go of the wall to go for bracing, nails, hammers, or to help others. Invariably, the wall topples over the floor onto the ground. Avoid this by delegating tasks before you lift the wall.

Once erected to an approximate vertical position, position the studs over their respective marks on the shoe. While one or more people hold the wall, the others must toe-nail the studs in place. Once toe-nailed, bracing made from stock 2 x 4s 12 ft long or longer is fastened on studs with 16d common nails driven partway home. The other end is fastened with nails to an adjacent shoe or a separate block nailed to the deck (Figure 7-6).

A real aid in erecting walls is to *plumb* each complete wall as it is erected. To plumb a wall requires the use of a level. Either end bubble on the level may be used to plumb a wall (Figure 7-7). (The center bub-

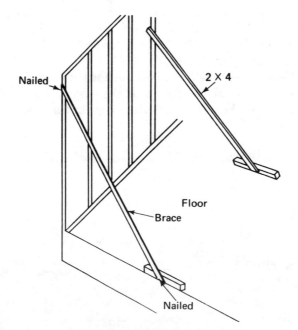

Figure 7-6 Bracing a Wall

Figure 7-7 Level   (*Courtesy of Stanley Tools*)

ble is used for *leveling.*) When the upper bubble is between the hairlines on the curved glass cylinder, the plumb condition is achieved. Simply hold the level side against the edge of the 2 x 4 stud near the brace and adjust the wall *in and out* until the wall is plumb. Fasten the brace to the floor or adjacent shoe. Repeat this process until the full length of the wall is plumb.

Reposition the level to the flat side of the stud and plumb the wall *left to right.* Use another stud to brace the wall when it is plumb.

Continue building walls and erecting them until they are all complete. Tie all walls together with the plate at the corners and at splices.

DOUBLE-PLATING THE WALL

Double-plating some walls is required by building codes. It is important to understand the need for double plates. A double plate is what its name implies: a second 2 x 4 on top of the first plate. Its purpose is to strengthen the walls. Therefore, each corner will be cross-braced and each splice will be offset. What do these ideas mean? Cross-bracing a corner is explained with the help of Figure 7-8.

Figure 7-8A shows how a finished outside corner looks *after* the second plate has been installed. Note that the first and second plates are placed so that joints labeled 1 and 2 are offset. In this manner the corner is strengthened.

Figure 7-8B shows how installation of a double plate is made where an intermediate wall intersects another wall. Notice that the double plate

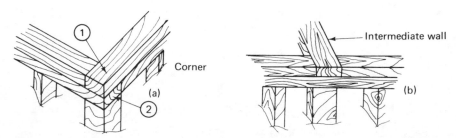

Figure 7-8 Cross-Bracing Corners

over the intermediate wall extends even with the outside of the second wall. Note, too, that a double plate is butted against each side of the intermediate's double plate. You can now double-plate each wall in your framework.

Two more tasks need to be accomplished before the job of framing walls can be called complete. They are: (1) making and installing inside corners, and (2) bracing walls. Each is essential for its own reasons.

MAKING AND INSTALLING INSIDE CORNERS

Inside corners are made in various ways. They are made with common studs and with 1-in. stock with the help of 2 x 4 cats. A *cat* is a piece of 2 x 4 or other stock lumber nailed between two studs. Cats function as braces, backing pieces for corner boards, and as fire breaks in walls.

Once again it is desirable to use an illustration to explain different ways of making and installing inside corners. Figure 7-9 shows a few views.

In Figure 7-9A a piece of 1 x 8 8 ft long is centered and nailed to the back side of the last stud on the intermediate wall. Then three 2 x 4 cats are cut to fit between the common studs. These are positioned one near the floor, one near the plate, and one near the midpoint between the plate and the shoe. They are nailed to the back side of the 1 x 8, then through the common studs into the ends of each cat.

Figure 7-9B shows that an inside corner needs to be made on one side of the intermediate wall. The other side happens to have a common stud in the correct position. Shown is another common stud nailed flat side to flat side to create the inside corner. If the nails are driven at 12 to 16 in. intervals and toe-nailed at plate and shoe, the corner will stay secure. Cats may be added as shown in A if more rigidity is needed or desired.

Figure 7-9C is a variation of B. In this case either one or two additional common studs are installed on each side of the last intermediate stud. Three blocks of 2 x 4 3-1/2 in. long are cut and installed in the pocket created by the three studs. Each 2 x 4 stud is nailed to the blocks. This stiffens the corner. Select the type of inside corner that best suits your needs. You will find that Routine BF5 provides you with detailed instructions, material requirements, and tools needed to accomplish this phase of wall construction.

CUTTING AND INSTALLING PERMANENT BRACES

The final task in constructing walls is to cut in and nail permanent braces. Although two methods are available, only one is developed in

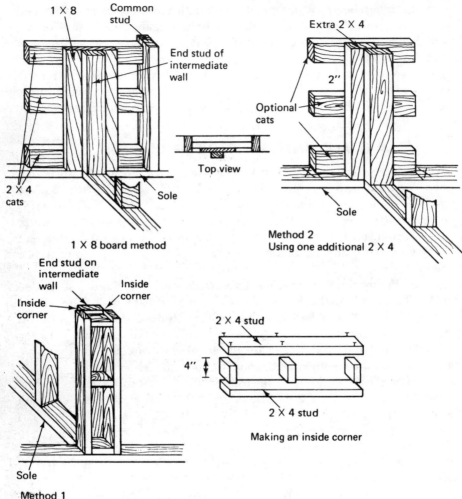

Figure 7-9 Inside Corners

Routine BF7. One method (not detailed) requires the insertion of a 1 x 4 into the common studs. This is done by first marking the position of the brace. Next, saw kerfs are made to a 3/4-in. depth. Following that, a chisel and hammer are used to chip out the stock between cuts, creating a dado. Finally the 1 x 4 is fitted into the dadoes and is nailed into place.

The second method shown in Figure 7-10 makes use of 2 x 4 stock. The 2 x 4s are cut on a bench and nailed between the studs. View Figure 7-10, and note that a line has been struck from the plate and corner area to the floor area. This is the relative position for bracing each outside

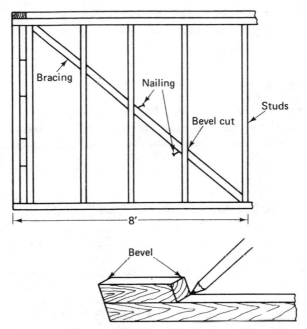

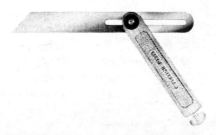

Figure 7-10 Corner Bracing

corner in the framework. You may cut and install the bracing from either
the plate down or the shoe up; it makes no difference. What is important
is to be sure that the wall has not moved and that it is *plumb* before you
start. A tool that reduces the difficulty of marking corner braces on a
bevel is called a *bevel square* (Figure 7-11). The wing nut loosens the
steel blade, allowing the blade to be rotated to any angle. After you have
marked the first brace on its bevel by holding it against the line, the bevel
square should be set. This bevel is then used to mark each succeeding
piece of 2 x 4 bracing. After marking the brace with the bevel square,

Figure 7-11 Bevel Square   (*Courtesy of Stanley Tools*) use a combination square to draw a line across the flat side of the 2 x 4 brace stock. Repeat this procedure until all pieces are fitted and installed.

ROUTINES

Now that you have read about it and realize that your only obstacle is work, embark on an enjoyable part of building—framing. The eight routines that follow cover a great variety of wall construction procedures.

BF1: WALL LAYOUT AND CONSTRUCTION

RESOURCES

Materials:
2 × _____ lineal ft of 2 x 4 = to length of wall _____ bd ft
3 2 x 4s 8 ft long for each 4 ft of wall length _____
1 lb 12d or 16d common nails per 8 ft of wall _____
1/2 lb 8d common nails per stud for toe nailing _____
2 2 x 4s 12 ft long for bracing (may be used elsewhere)

Tools:
1 no. 8 crosscut handsaw
1 framing square
1 50-ft tape or 6-ft wooden ruler
1 16-oz claw hammer
1 pair sawhorses
1 24-in. level

ESTIMATED MANHOURS

45 minutes per 8 ft of wall

PROCEDURE

Step 1 Lay the shoe (sometimes called the sole) and the plate along the position where the wall will be installed (Figure 7-2A).

Step 2 Mark with a square the first stud position at 15-1/4 in. from the end.

Step 3 Mark each additional stud location 16 in. from the first. Make an **X** on the same side of each 16-in. line.

Step 4 Trim or cut the shoe and the plate at the end of the wall area.

Step 5 On the sawhorses, lay out and cut all the studs needed.

If a door or window is included in the wall, proceed with the next instructions; if not, skip to instruction 9.

Step 6 Lay out the window center from a measurement taken from the plan.

Step 7 Measure the window width from the window center line as shown on the plan (one-half of the width to each side of the center line).

Step 8 Repeat steps 6 and 7 for the door opening.

Step 9 Lay a 2 x 4 where every mark was made on the shoe and the plate. Do *not* place studs within window or door openings.

Step 10 Nail the shoe in place.

Step 11 Nail the plate to the top of the studs with 12d or 16d common nails.

Step 12 Nail in place the window-opening unit constructed in Routine BF3 and/or door-opening unit constructed in Routine BF4 (if applicable).

Step 13 Nail in place the corner posts constructed in Routine BF2 (if applicable).

Step 14 Raise the wall and *toe-nail* each stud in place with four 8d common nails per stud.

Step 15 Plumb the wall with the level and secure with a brace.

BF2: PREPARING OUTSIDE CORNER POSTS

RESOURCES

Materials:
  3 2 x 4s 8 ft long
  3 blocks of 2 x 4 15-7/8 in. long
  27 12d common nails

Tools:
1 no. 8 crosscut handsaw
1 16-oz claw hammer
1 combination or framing square
1 pair sawhorses

ESTIMATED MANHOURS

15 minutes

PROCEDURE

Step 1   Determine the total length of the studs from the plan.

Step 2   Mark the length of a stud after squaring the end with a saw.

Step 3   Cut three 2 x 4s to length needed.

Step 4   Cut three blocks of 2 x 4 15-7/8 in. long.

Step 5   Nail the blocks to the flat side of a 2 x 4.

Step 6   Nail the flat side of a second 2 x 4 to the blocks.

Step 7   Nail the flat side of a third 2 x 4 to the edge of the first 2 x 4 and the blocks.

BF3: CONSTRUCTING A WINDOW-FRAME UNIT

RESOURCES

Materials:
2 2 x 4s 8 ft long (cut to the actual required stud length)
24 to 32 lineal ft of 2 x 4 stock for windows up to 48 in. (add 6 to 8 ft for each additional 16 in. of window width)
2 2 x 6s, 2 x 8s, 2 x 10s, or 2 x 12s x L (header) (for L, see Tables BF3A and BF3B at the end of the procedure section)
1 2 x 4 x L (sill)
1/2 lb 12d common nails

Tools:
1 no. 8 crosscut handsaw
1 framing square
1 pair sawhorses
1 6-ft folding ruler

ESTIMATED MANHOURS

30 minutes for windows up to 48 in. wide
add 10 minutes per additional foot of width

PROCEDURE

Step   1   Cut two 2 x 4s the full length of a common stud.

Step   2   Cut two pieces for headers.

Step   3   Cut three or more spacers (as required) 1-1/2 in. x 1-1/2 in. W, where W = the width of the header (e.g., for a 2 x 6, cut W 5-1/2 in. long).

Step   4   Nail the spacers to one header.

Step   5   Nail the headers together with 12d common nails.

Step   6   Cut the sill the same length as the header.

Step   7   Mark stud placement on the sill according to marks made on the shoe. Then transfer another set to the top of the header.

Step   8   Mark the height of the window from the floor on full-length studs. (Use your plan to obtain this measurement.)

Step   9   Measure and mark the top of the window opening. (Again, use your plan to obtain this measurement.)

Step 10   Nail the sill *below* (*but even*) with the lower mark on the studs.

Step 11   Nail the header *above* (*but even*) with the upper mark on the the studs.

Step 12   Cut a lower jack stud:
a. To nail on each full-length stud.
b. To nail in each 16-in. required place.

Step 13   Nail these 2 x 4s into place with 12d common nails and 8d common nails (where toe nailing is needed).

Step 14   Measure and cut an equal number of upper jack studs and nail into place.

Step 15　Measure and cut two middle jack studs and nail into place within the opening.

Step 16　Finally, lift the unit and place it in the wall section before lifting the wall. Nail.

TABLE BF3A: Headers for Non-Load-Bearing Walls

Span (ft)	Lumber Size
Less than 3	Two 2 x 4s on edge
3–5	Two 2 x 6s on edge
5–8	Two 2 x 8s on edge
8–12	Two 2 x 10s on edge
Over 12	Two 2 x 12s on edge

TABLE BF3B: Headers for Load-Bearing Walls

Span (ft)	Lumber Size
Less than 3	Two 2 x 6s on edge
4–6	Two 2 x 8s on edge
6–10	Two 2 x 10s on edge
10–16	Two 2 x 12s on edge

BF4: CONSTRUCTING A DOOR-WALL UNIT

RESOURCES

Materials:
2　2 x 4s 8 ft long for common studs
20　ft of 2 x 4s for jack studs

2 x 6 x L headers (see Table BF4B for header requirements for doors over 4-ft wide)

1/2 lb 12d common nails

1/4 lb 8d common nails

3 to 8-1/2 in. x 1-1/2 in. x 5-1/2 in. spacer pieces

Tools:
1 no. 8 crosscut handsaw

1 16-oz claw hammer

1 framing square

1 pair sawhorses

1 6-ft folding ruler

ESTIMATED MANHOURS

20 minutes

add 3 minutes per foot over 4 ft

PROCEDURE

Step 1 Cut two common studs per measurement on the plan.

Step 2 Cut two pieces for headers per length listed in Table BF4A.

Step 3 Cut four or more 1/2-in. spacers.

Step 4 Nail the header together as per Figure BF4A.

Step 5 Measure 6 ft 9 in. from the end of the common studs and make a line.

Step 6 Position the header next to the shoe on the floor and transfer the stud placement.

Step 7 Nail the header to the common studs by placing the header above the marks on the studs.

Step 8 Cut two jack studs 6 ft 9 in. and nail under the header and to the common studs.

Step 9 Cut one upper jack stud for each 16 in. of stud requirement + two jack studs to nail to the common studs.

Step 10 Position the door-wall unit in place and nail to the plate and shoe.

TABLE BF4A: Door-Header Lengths for Standard Doors

Door Width	Header Lengths	
	Interior Doors	Exterior Doors
20 in.	25-1/4 in.	26-1/2 in.
24 in.	29-1/4 in.	30-1/2 in.
26 in.	31-1/4 in.	32-1/2 in.
30 in.	35-1/4 in.	36-1/2 in.
32 in.	37-1/4 in.	38-1/2 in.
36 in.	41-1/4 in.	42-1/2 in.
40 in.	45-1/4 in.	46-1/2 in.
48 in.	53-1/4 in.	54-1/2 in.
5 ft (double doors)	5 ft 5-1/4 in.	5 ft 6-1/2 in.
6 ft (double doors)	6 ft 5-1/4 in.	6 ft 6-1/2 in.
7 ft (double doors)	7 ft 5-1/4 in.	7 ft 6-1/2 in.
8 ft (double doors)	8 ft 5-1/4 in.	8 ft 6-1/2 in.

Interior door jambs are 3/4 in. thick.
Exterior door jambs are 1-1/4 to 1-3/8 in. thick.

TABLE BF4B: Header Size for Door-Wall Units (Non-Load-Bearing Wall)

Up to 48-in. doors, use 2 x 4s or 2 x 6s.
48 in. to 8 ft, use 2 x 8s.
8 ft to 12 ft, use 2 x 10s.

BF5: CONSTRUCTING INSIDE CORNERS

RESOURCES

Materials:
1 x 8 board method:
   1   1 x 8 8 ft long
   3   pieces of 2 x 4 15-1/4 in. or 23-1/4 in. long (depending on stud centering—16 or 24 in. OC)

Additional 2 x 4 methods:
    1 or 2 2 x 4 8 ft long
    6 to 8 12d common nails
    1/2 lb 8d common nails

Tools:
1 no. 8 crosscut handsaw
1 16-oz claw hammer
1 combination square
1 pair sawhorses

ESTIMATED MANHOURS

15 to 20 minutes

PROCEDURE A: 1 x 8 BOARD METHOD

Step 1   Cut a 1 x 8 8 ft long to the common-stud length.

Step 2   Center the board on the back of a 2 x 4 stud and nail with 8d common nails.

Step 3   Cut three braces equal in length to the spacing between the studs.

Step 4   Nail the braces one on the top, one on the bottom, and one in the center to the back of the 1 x 8.

Step 5   Nail the 2 x 4 braces to the adjacent 2 x 4 common studs.

Tip: Take a scrap piece of 1 x 8 and use it as a guide while nailing the center brace to its 2 x 4s. When completed, the 1 x 8 will be aligned with the common wall studs and will make a straight corner.

PROCEDURE B: ADDITIONAL 2 x 4 METHOD 1

Step 1   Cut two additional 2 x 4s to the common stud length.

Step 2   Cut three pieces of 2 x 4 4 in. long.

Step 3   Nail 2 x 4 studs to 4-in. pieces.

Step 4   Position the assembly and toe-nail with 8d common nails.

Step 5   Toe-nail the third stud to three pieces or drive 12d common nails through the third stud into three pieces.

PROCEDURE C: ADDITIONAL 2 x 4 METHOD 2

Step 1   Cut one 2 x 4 to the common stud length.

Step 2   Nail the stud to a second stud with 12d common nails and into the shoe and plate with either 8d or 12d common nails.

PROCEDURE D: BENCH METHOD

Note: Any one of the three methods detailed above may be made at the bench except for nailing the bracing. In each case the previously installed stud would be a part of the bench construction.

BF6: ERECTING AND PLUMBING WALLS

RESOURCES

Materials:
2 x 4 12 to 16 ft long (varying in length)
2 x 4 scraps 2 ft long for nailing to floor unless all shoes have been
   installed
16d common nails (for brace nailing)
8d common nails (for toe nailing)

Tools:
1 16-oz claw hammer
1 pair sawhorses
1 24-in. level

ESTIMATED MANHOURS

20 to 30 minutes per 12 ft of wall

PROCEDURE

Step 1   Butt previously assembled wall section to previously installed shoe.

Step 2   With two people, raise the plate and studs erect.

Step 3   Place the studs on the shoe. (See Tip 1.)

Step 4   One person tack-nails a brace near the plate on the stud while the second person holds the wall vertical. (See Tip 2.)

Step 5   Approximate a vertical wall position and tack-nail the bottom of the brace to an available shoe on an adjacent wall.

Step 6   Repeat steps 4 and 5 to install the other braces. (Usually two are sufficient for walls up to 16 ft long. Over 16 ft, use a brace somewhere near the wall splice.)

Step 7   Plumb the wall for in—out vertical with the level each place a brace is used. Release the bottom end of the brace; reposition and renail.

Step 8   Plumb the wall in line by placing the level on the flat side of a stud, and, using another 2 x 4 brace, tack-nail in position.

Step 9   Raise the other walls and join where the corners meet.

Tip 1: For heavy sections of walls, partially raise the wall section and slide the sawhorses under the plate. Rest a moment, then raise the wall with the muscles of your upper legs, back, and shoulders. This method may prevent injury.

Tip 2: Decide which person will hold the wall. If both people let go to get braces, you can expect to see the wall disappear over the side. (I've never seen a wall fall back onto the floor, always over the side.)

BF7: INSTALLING CORNER BRACING

RESOURCES

Materials:
2 2 x 4s 12 ft long for each wall section
1 lb 8d common nails per two pieces of 2 x 4

Tools:
1 no. 8 crosscut handsaw
1 chalk line
1 bevel square
1 combination square
1 pair sawhorses
1 16-oz claw hammer

ESTIMATED MANHOURS

30 minutes per brace

PROCEDURE

Step 1 Snap a chalk line from the plate corner to a point along the floor 8 ft in from the corner.

Step 2 Position a 2 x 4 along the chalk line and mark the 2 x 4 where it touches the first stud. Also mark it on the shoe.

Step 3 Square the bevel mark with the combination square.

Step 4 Cut both ends.

Step 5 Install the piece using 8d nails.

Step 6 Set the bevel square for the angle present on the long part of the remaining 2 x 4.

Step 7 Position the 2 x 4 along the chalk line again and mark alongside the next studs.

Step 8 Square and cut the piece.

Step 9 Using the piece just cut, lay out enough pieces to equal one brace piece between each stud. Square and cut.

Step 10 Install all brace pieces cut.

Step 11 Hold 2 x 4 stock against the chalk line and mark the final top piece. Cut and nail into place.

BF8: DOUBLE-PLATE INSTALLATION

RESOURCES

Materials:
2 x 4 x L, where L equals the total lineal feet of wall
3 to 5 lb 12d or 16d common nails

Tools:
1 no. 8 crosscut handsaw
1 pair sawhorses
1 6- or 8-ft stepladder
1 16-oz hammer

ESTIMATED MANHOURS

15 minutes per 8 ft of wall

PROCEDURE

Note: You can start anywhere, but frequently the outside walls are doubled first.

Step 1 Lay the first 2 x 4 on top of the plate so that the outside corner is covered by this 2 x 4.

Step 2 Cut its length equal to where the inside wall joins the outside wall.

Step 3 Nail in place with 12d or 16d common nails.

Step 4 Put a scrap piece of 2 x 4 over the joint. Butt the double plate 2 x 4 against the scrap piece.

Step 5 Cut and nail the second piece of double plate as required.

Step 6 Remove the scrap piece between the first and second double plates and all others. Fit, cut, and install all inside walls with double plates.

8

JOIST FRAMING

BASIC TERMS

Box beam member of a joist foundation of the same proportions as the other joists, placed at the ends of a row of joists and nailed to those joists' ends; sometimes called an **L**-*type sill.*
Bridging method of reinforcing a joist foundation to make it more solid and free of movement. *Cross-bridging* and *solid bridging* are the two most common installation methods.
Cleat piece of stock usually 5/4 in. x 3 in. or 2 x 4 in. nailed to a member that aids in supporting another member which will partially rest upon it.
Crown edge of a joist where the bow in the joist seems to be *up* in the middle of its length.
Dead load term used to account for the weight of building materials ap-

plied to a floor or ceiling (e.g., wallboard, plaster, subflooring, finished flooring).
Joist structural member stood on edge and laid horizontally to rest on some points; usually made from 2-in. stock in widths ranging from 4 to 12 in. and lengths up to 16 ft and longer.
Live load weight of people and furniture that will be extensively applied to the floor.
Sill installed 2-in. stock member, usually found on top of a cement foundation or wall; the member to which joists are nailed.
Span distance between two points on which a joist will be laid to rest. The span can be measured from sill to opposite sill or plate to opposite plate.

Joist framing is the part of construction that provides the foundation for the floor or the foundation from which the ceiling will hang. In two-story construction, a single joist unit installed between the first and second floors acts as both floor and ceiling foundation. Flooring is nailed to the top edge of the joist, and ceiling material is fastened to the bottom edge of the joist.

SCOPE OF THE WORK

The scope of work involved in installing a joist foundation is rather extensive. Considerable time must be dedicated to its planning and preparation. This includes laying out sills and installing them onto foundations, placing joists, and cutting each joist. Additional time must be dedicated to the actual installation of the joists.

Following this work, additional time will be spent in aligning and reinforcing the joists. The reinforcement involves preparing and installing the bridging.

In many applications structural members such as 2 x 10 and 2 x 12 joists are used. When members of these sizes are used, it is a good idea to have at least one more person available, to aid in lifting, positioning, and nailing.

The previous paragraphs outlined the bare essentials necessary to install joist units. Variations are frequently encountered that require considerable study and adaptation to the basics. Two of these are: (1) adding a room to a house using cleating (explained in detail later), and (2) tying in to the present ceiling or floor joists (also explained in detail later in this chapter). In each of these variations there are many more problems to solve. However, the primary objective to achieve is the matching with the old work.

Each of the variations in this task must be independently studied and resolved. When any variation is expected, exposed, and planned for, a certain amount of time and material should be added to the requirement list of your bill of materials. In addition, a comment about the situation should be made in your programmed plan description section. This will alert you to details about this area of work.

Joists for the floor or for the ceiling are basically the same but have two differences. Floor joists must carry a live load (people, furniture), whereas ceiling joists usually carry only a dead load (wallboard and insulation). The exception is where the ceiling joists form a part of the roof or are the floor joists for the second floor.

LIVE AND DEAD LOADS

In each application joists are spaced either 16 or 24 in. OC, in accordance with building codes and needs. Tables 8-1 and 8-2 list the live-load capability of floor joists for various purposes. Maximum spans are shown for joists spaced at 16 and 24 in. OC. This translates to a special size of structural member for a specific span. The span is the distance between points where the ends of the joist rest (Figure 8-1).

Table 8-3 lists various sizes of stock lumber that can be used for ceiling joists. Note that the table calls for dry wall ceiling materials and limited storage on the ceiling. If a room is made upstairs sometime in the future, joists must then be selected and used that will carry the appropriate loads.

TABLE 8-1: Live-Load Span for Floor Joists (40 psf) for All Rooms Except Sleeping Rooms and Attic Floors

| Size of Stock | Inches on Center | Lumber Grade | |
		No. 1 Common	No. 2 Common
2 x 6	16	10-4	9-6
	24	9-0	8-1
2 x 8	16	13-7	12-7
	24	11-11	10-8
2 x 10	16	17-4	16-0
	24	15-2	13-7
2 x 12	24	21-1	19-6
	16	18-5	16-6

Based upon a live load of 40 lb/sq ft. The lengths listed under the two grades of lumber are in feet and inches of span. Both grades are usually available at local lumberyards. Utility grades of lumber should *never* be used for live-load applications.

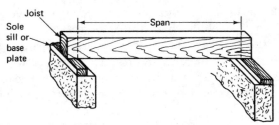

Figure 8-1 Joist and Span

Let's study these tables by use of an example. Assume that a new addition is being added to your home. Assume further that its width is 12 ft and its length is 14 ft. It is to be used as a family room. Table 8-1 must be used for this. If the joists are spaced 16 in. OC and span the 12-ft width, and no. 2 common grade is used, Table 8-1 shows that 2 x 8s must be used. If the joists are spaced 16 in. OC and span the 14-ft length, 2 x 10s must be used.

CROWN UP

As the trees bend with the wind, so do joists bend when spanning areas and when live and dead loads are placed upon them. Therefore,.

it is very important that the joist be placed on its sills in a manner that takes advantage of the natural strength of the beam. The best way to do this is by placing the beam with its *crown up*.

TABLE 8-2: Live-Load Span for Floor Joists (30 psf) for Sleeping Rooms and Attics

Size of Stock	Inches on Center	Lumber Grade	
		No. 1 Common	No. 2 Common
2 x 6	16	11-4	10-6
	24	9-11	9-0
2 x 8	16	15-0	13-10
	24	13-1	11-11
2 x 10	16	19–1	17-8
	24	16-8	15-3
2 x 12	16	23-3	21-6
	24	20-3	18-6

Based upon a live load of 30 lb/sq ft. The lengths listed under the two grades of lumber are in feet and inches of span.

Table 8-3: Live-Load Ceiling Joists (20 psf) for Drywall Ceilings with Limited Storage on Ceiling Joists

Size of Stock	Inches on Center	Lumber Grade	
		No. 1 Common	No. 2 Common
2 x 4	16	9-6	8-7
	24	8-3	7-0 *
2 x 6	16	14-11	12-3 *
	24	13-0	10-0 *
2 x 8	16	19-7	16-2 *
	24	17-2	13-2 *
2 x 10	16	25-0	20-7 *
	24	21-10	16-10 *

Based upon ceiling joists with a live load of 20 lb/sq ft. The lengths marked with an asterisk are based upon maximum stress of fiber and have a 10 lb/sq ft dead-load factor included. If spans of these lengths are in your plans, it would be best to increase the size of the stock or to select a no. 1 common grade.

In order to determine where the crown is, a builder must *sight* the beam. To sight a beam means to stand a beam on its edge while holding one end aloft and then with your eye, sight to determine its bend (Figure 8-2). If the beam *sags*, the crown is on the bottom. If the beam bows up or looks straight, the crown is up. Make a mark with a crayon on the crown side of the beam. Mark all beams that will be used with an identifying X on the crown edge.

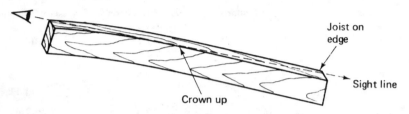

Joist on edge

Sight line

Crown up

Figure 8-2 Sighting a Beam

Why is determining the crown necessary? If joists are installed with some crowns up and others down, the floor or ceiling will be uneven. In floor construction, this situation will cause a squeaky floor. Subflooring will pull loose from beams that are installed with the crown down. The loosened flooring and nails will lift the finished floor, making it loose.

If ceiling joists are installed with the crown down, the effect of dead loads will cause a greater sag. The result will be that fixtures and installed materials will begin to tilt. The sag will become progressively worse until the beam has reached its maximum fiber stress point and is completely dried. With all the crowns up, sagging will still take place but will usually stop as the beams reach a straight-line plane.

Three detailed applications are discussed: (1) the box sill/joist construction method, (2) the cleating method for additions, and (3) the tying-in method.

BOX SILL/JOIST CONSTRUCTION

The box sill/joist construction method is usually associated with floor foundation building. It is almost never used with ceiling foundations except in two-story homes, where the ceiling and floor joists serve a dual function. Figure 8-3 illustrates the finished box sill/joist construction of a floor foundation. The parts that make up the assembly are:

1. Sill
2. Box beam
3. Joists
4. Bridging

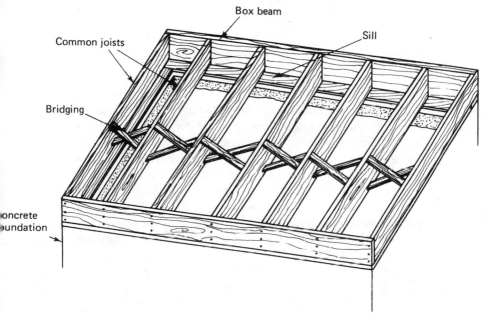

Figure 8-3 Box Sill/Joist Construction

The work involved in building this type of foundation requires a few specific individual tasks. First, the sills must be cut and installed. Since the joists will set upon a sill, the sill will need to be wide enough to provide a good seat. It is customary to provide from 4 to 6 in. of seat for each joist end. In the box sill/joist construction method an additional 1-1/2 in. width must be added to the width of the sill to account for the box beams (Figure 8-1 or 8-3). The sills are usually bolted to the cement foundation where long bolts have previously been set in cement. (Refer to Chapter 6 for a review if needed.) Position the sill against the bolts sticking from the cement and mark each bolt's position. Next measure in from the outer-edge reference point to the center of the bolt as shown in Figure 8-4. Transfer the distance measured to the sill and drill a hole large enough to allow passage of the bolt. Place the sill in position, install a washer, and tighten the sill to the foundation with nuts. Repeat this process until all sills are in place. Toe-nail all places where the sills join each other. This will be at splices on long walls and at corners.

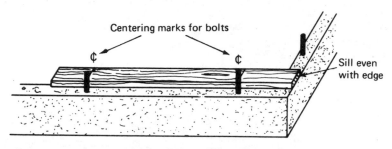

Centering marks for bolts

Sill even with edge

Figure 8-4 Marking for Bolts

Following the installation of the sill, the box beams are cut and installed. A member or more than one member may be required to make one box-beam length. This member is usually toe-nailed in position. Its outer edge usually is flush with the sill. Once installed the joist layout is made. The layout consists of marking each joist's position on both the sill and the box beam. In some cases carpenters just mark the box beam. But then they square a line from the top of the box beam down to the sill on the inside of the beam. Using this method, they feel, aids them in actually installing the joists. Place an X on the side of the mark where the beam will be placed.

Cut all joists of a common length at one time. Before cutting each beam it should be sighted for *crown*. Once the crown is determined, all measuring and squaring should be made from the crown edge. This method will ensure more reliable fitting when joists are installed.

After all the joists have been positioned (with crown *up*) and have been toe-nailed to the sill and face-nailed through the box beam, they must be spaced throughout their length. For this job a space board made from a 1 x 4 is used. The 1 x 4 is marked at 16 or 24-in.-OC intervals as required (Figure 8-5). Its purpose is to provide: (1) a guide for spacing the joists, and (2) a fastening block to keep the joists in position until the bridging is installed. Place the spacer board 90 degrees to the run of the joists and 12 to 15 in. from chalk lines where each row of bridging will be installed. This method assures that the bridging will be installed properly. Figure 8-6 shows what a chalk line looks like.

Where do you make the lines for bridging? First, bridging is needed

1 X 4   16" OC or 24" OC

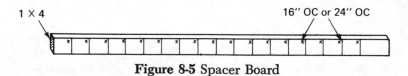

Figure 8-5 Spacer Board

Figure 8-6 Chalk Line *(Courtesy of Stanley Tools)*

in each joist foundation to provide stiffness. Without it, persons walking upon a floor, for example, will have a sense of swaying and sinking. As a rule of thumb, a row of bridging is installed 6 to 8 ft across a joist foundation. In a 12-ft span one set of bridging will usually be sufficient. In a 16-ft span, one set of bridging will be required and there may be a need for two sets.

Bridging can be installed using different materials and methods. The solid bridging method requires that pieces of joist members be cut in lengths equal to the spacing between the joists. For 16 in. OC these would be 14-1/2 in. long except for the first and last spacing. The first and last pieces must be custom cut. Figure 8-7 shows the three methods of bridging: A illustrates solid bridging. Notice that each piece of

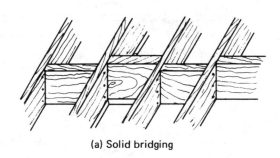

(a) Solid bridging

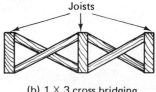

(b) 1 × 3 cross bridging

(c) Steel cross bridging

Figure 8-7 Bridging Methods

bridging is nailed on the opposite side of the chalked line. This is done to make installation easier. Size 16d common nails should be driven through the joist and into the bridging piece.

In Figure 8-7B, bridging made from 1 x 3 no. 2 common lumber is precut at angles and toe-nailed to both the top and bottom of opposite joists. This method is known as *cross-bridging*. Table BJF3 (included in Routine BJF3, Cross-Bridging) provides you with settings of the framing square that will give the correct angles to cut bridging. To understand how to use the table, examine Figure 8-8.

The 1 x 3 stock is stood on edge and the framing square is posi-

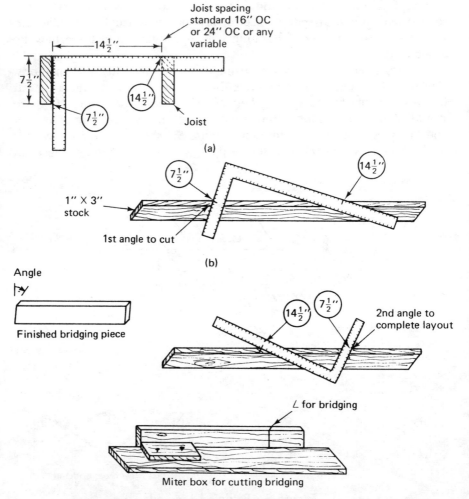

Figure 8-8 Laying Out Bridging with a Framing Square

tioned as though it were installed in the joist assembly (Figure 8-8A). The two measurements that will be used are: (1) the width of the joists on the tongue of the square, and (2) the distance between the joists used on the blade of the square. In A these positions are 7-1/2 in. on the tongue and 14-1/2 in. on the blade. The two points are located along the outer edge of the 1 x 3 stock (Figure 8-8B) and a pencil mark is made on the outer edge of the tongue. The square is reversed and the 14-1/2-in. mark on the blake is aligned with the mark previously drawn (again on the outer edge of the 1 x 3). The 7-1/2-in. mark on the tongue and outer edge of the 1 x 3 are aligned. A second line is drawn on the 1 x 3 along the outer edge of the tongue. The resultant piece looks as shown in Figure 8-8. With one piece cut and tried, it can be used as a pattern to lay out the remaining pieces needed; or a miter box can be made. Figure 8-8 shows what a miter box would look like for cutting standard bridging pieces. Only one cut needs to be made in the miter box. Its angle can be determined by using the method shown in B of the figure or by using data from Table BJF3.

After all joists are nailed on top, they are nailed on the bottom. Because joists are sometimes twisted, some manual adjustment may be needed. While warping the joist in position with one hand, nail the bridging with the free hand or have someone help you with this.

The third method of bridging, shown in Figure 8-7C is similiar to the cross-bridging just detailed, except that steel bridging pieces are purchased and nailed to the joists.

CREATING METHOD FOR AN ADDITION

The first problem to overcome after exposing an old joist foundation is to determine how the new floor or ceiling joists will be supported. Should you find a box sill beam or a joist as the exposed beam where the new joists will be fastened, the simplest method for providing support is by installing a cleat (Figure 8-9).

A cleat made from either a piece of 5/4 in. x 3 in. or 2 x 4 stock is nailed to the lower edge of an existing joist. For 2 x 6, 2 x 8, and 2 x 10 joists, a 5/4 in. x 3 in. cleat is adequate. For 2 x 12 and some installations of 2 x 10 (long spans, and excessive live-load requirements), the 2 x 4 cleat is desirable. The cleat should be nailed at close intervals, 10 to 12 in. spacing, and distributed throughout the width of the stock not concentrated at the top, center, or lower edge of the cleat.

After the cleat(s) is installed, an *exact* measurement of the height from the top of the cleat to the floor must be made. Subtracted from this length are the thicknesses of the subfloor and finished floor which you plan to install. The remainder is the distance from the top of the joist

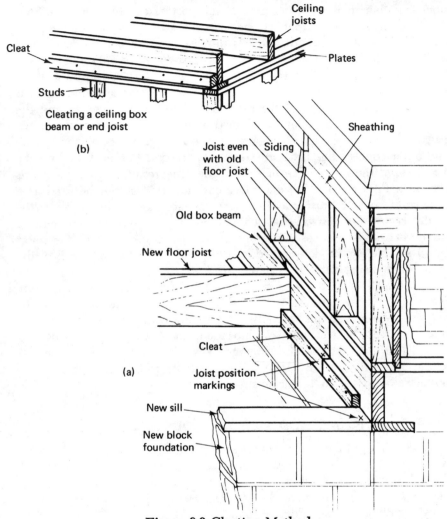

Figure 8-9 Cleating Method

to the top of the cleat cutout (Figure 8-9). Cut each joist in a similar manner.

Unless it is possible to physically nail through the old joist, the toe-nail method must be used. First lay out the position in which each joist will be installed. Toe-nail each joist in place with 10d or 12d common nails. Use at least 5 or 6 nails per joist, one on top and two or more on each side.

The opposite ends of the joist may either be boxed or left open, depending upon individual requirements. Install bridging as required.

TYING-IN METHOD

The tying-in method of joist foundation is similar to the cleating method except that when the old work is removed, the ends of the old joist are exposed. The old plates and/or sills are readily available for use.

To tie in to an old installation, each new joist should take full advantage of the available sill or plate. The length of the joist must be calculated from the inside edge of the old plate to the outer edge of the new plate.

Once again, since new wood is added to old and their dimensions differ, close measurements must be taken. Figure 8-10 shows what some of these measurements involve. The new joists *probably* are slightly smaller in size than the old; therefore, one of two decisions must be made: use joists of the same general size as the old ones (2 x 6, 2 x 8, etc.), or use the next larger size and custom-fit each joist.

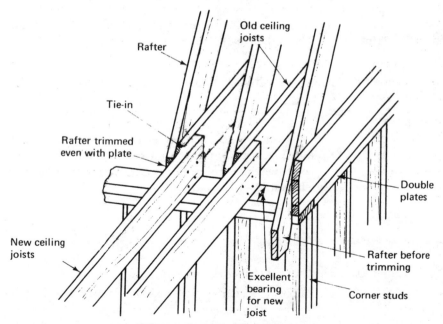

Figure 8-10 Tying-in Joists

If you elect the first solution the joist will need to be shimmed so that the top edge of the new joist is properly set. Shimming can be done very effectively with cedar shingles. If you decide on the second solution and use the next-larger-size joist, a cutout similiar to that made for cleating will probably be made. This method may be desirable for floor-joist application but will require additional work on both ends of the joist where it or they rest upon sills. If the larger joist is planned to be used, its size should be included in designing the outer edge sill or plate height.

Unless circumstances prevent it, plan to tie in the new beams by nailing them to the side of the old ones. This makes the one end really secure, perpendicular, and solidly installed. Lay out the opposite end according to the old joist layout. One exception to this plan would be where the new joists are to be spaced differently from the old ones. Avoid this problem if possible.

ROUTINES

The four routines that follow provide data on how to plan and build a floor or ceiling joist foundation. They define each phase of the various types of installations discussed in this chapter. Since each identifies only its basic requirements, you must use your plans to fill in the required data. These data will deal with such details as:

1. Spacing: 16 or 24 in. OC
2. Size of stock and length
3. Type of bridging
4. Type of installation

If you are building over the ground, these may be the first wood products you will install. If you are putting up a ceiling foundation, the walls are already up. In either case, enjoy this area of work; it is heavy, hard work, but really enjoyable.

BJF1: JOIST LAYOUT AND ASSEMBLY

RESOURCES

Materials:
1 1 x 4 x 16 in. for spacer board
L ft of 2 x 4, 2 x 6, or 2 x 8 sills for foundations and floor joists (none required for ceiling joists)

5 lb 8d common nails
Floor plan

Tools:
1 50-ft tape or 6-ft folding ruler
1 framing square
1 pair sawhorses
1 16-oz claw hammer
1 no. 8 crosscut handsaw

ESTIMATED MANHOURS

4 to 12 hours for rooms 12 × 12 ft

PROCEDURE

Step 1 Cut and install sills on the foundation (bolt in place).

Step 2 Mark the sill/plate for 16 or 24-in.-OC spacings according to your floor plan.

Step 3 Nail a cleat to the old foundation if adding an addition.

Step 4 Toe-nail box beam(s) in place if using box-beam construction.

Step 5 Position each precut joist in place, verify that the crown is up, and nail in place.

Step 6 Mark the spacer board, lay perpendicular to the joists, and tack-nail with 8d nails.

Step 7 Snap a line perpendicular to the joists' run for a layout of the bridging.

Step 8 Cut and install the bridging.

BJF2: JOIST LAYOUT AND CUTTING

RESOURCES

Materials:
_____ 2 x _____ x _____ joist members
   no.   size   length

_____ 2 x _____ x _____ box beam
   no.   **size**   **length**

Floor plan

Tools:
1 pair sawhorses
1 framing square
1 wooden folding ruler or 25-ft steel tape
1 no. 8 crosscut handsaw
1 combination square

ESTIMATED MANHOURS

15 minutes per joist

PROCEDURE

Step 1 Place a few structural members, each with crown heading in the same direction, across a pair of sawhorses.

Step 2 Place the blade of the square along the crown edge of the member.

Step 3 Mark a square line across one end of the member by using a framing square and pencil.

Step 4 Measure and mark the required length of the joist from the line just drawn toward the other end of the joist.

Step 5 Square the mark just made.

Step 6 Cut both ends with the handsaw.
Skip to step 8 if cleating is not used.

Step 7 If cleating is used, make a cutout for the cleat. Use Figure 8-9.
a. Measure the distance from the top of the old beam to the top of the cleat.
b. Draw a line in from the end at the point. Measure down from the top edge.
c. Draw a connecting line from the bottom edge up to the line in from the joist end equal to the thickness (5/4 or 1-1/2 in.) of the cleat.
d. Cut out a notch for a cleat.

Step 8 Try the first joist for fit.

Step 9 Mark and cut the remaining members, using the first joist as a pattern.

BJF3: CROSS-BRIDGING

RESOURCES

Materials:
$$\frac{\underline{\qquad\qquad}}{\text{L ft}} \text{ 1 x 3 stock for bridging}$$

2 to 5 lb 6d common nails

Tools:
1 pair sawhorses
1 no. 8 crosscut handsaw
1 bevel square
1 framing square
1 chalk line
1 16-oz claw hammer
1 handmade or adjustable miter box (optional)

ESTIMATED MANHOURS

1.5 hours for 12-ft run of bridging

PROCEDURE

Step 1   Use Table BJF3 or the method explained in the text to cut one pair of bridging pieces for each common space between joists.

Step 2   Measure in at two points 6 to 8 ft from a common reference point perpendicular to the joist run.

Step 3   Snap a chalk line between the points marked in step 2.

Step 4   Repeat steps 2 and 3 as often as necessary (one line for each row of bridging).

Step 5   Install a spacer board, previously marked, 15 in. from the chalk line.

Step 6   Prenail the bridging pieces with two 6d common nails at each **end.**

Step 7 Nail each piece of bridging even with the top of the joist. Within a space, nail one piece of bridging on each side of the chalk line.

Step 8 Cut a pair of bridging pieces for each nonstandard space. (Use the framing square to obtain the length and angle.)

Step 9 Install the nonstandard pieces according to steps 5 and 6.

TABLE BJF3: Cross-bridging   *Select from column 1 a size of stock to be used and a number of inches of joist spacing (on center). Lay out the bridge pieces according to columns 2 and 3. Set a miter box or cut a miter in a miter box for the angle shown in column 4. Estimate the lineal feet of bridging needed for the job by using column 5 × the number of spaces between joists.*

(1) Size of Joist (in.)	OC (in.)	(2) Framing Square Tongue (in.)	(3) Framing Square Blade (in.)	(4) Angle (degrees)	(5) Length of Stock per Set (in.)
2 x 4	12	3-1/2	10-1/2	71	22-1/4
	15-1/4		13-3/4	75	28-1/2
	16		14-1/2	77	30
	23-1/4		21-3/4	80	44-1/4
	24		22-1/2	81	46-1/4
2 x 6	12	5-1/2	10-1/2	62	24
	15-1/4		13-3/4	68	30
	16		14-1/2	69	31-1/2
	23-1/4		21-3/4	75	45
	24		22-1/2	77	46-1/2
2 x 8	12	7-1/2	10-1/2	55	26
	15-1/4		13-3/4	61	31-1/2
	16		14-1/2	62	33-1/2
	23-1/4		21-3/4	70	46-1/2
	24		22-1/2	72	47-1/2
2 x 10	12	9-1/4	10-1/2	50	28-1/2
	15-1/4		13-3/4	55	33-1/2
	16		14-1/2	57	35
	23-1/4		21-3/4	67	47-1/2
	24		22-1/2	68	49
2 x 12	12	11-1/4	10-1/2	45	31
	15-1/4		13-3/4	50	36
	16		14-1/2	53	37-1/2
	23-1/4		21-3/4	62	49-1/2
	24		22-1/2	63	51

BJF4: SOLID BRIDGING

RESOURCES

Materials:
$\underline{\hspace{3cm}}$ of 2 x $\underline{\hspace{3cm}}$ for bridging
$\quad$ L ft $\qquad\qquad$ size of joist

5 to 10 lb 12d or 16d common nails

Tools:
1 pair sawhorses
1 no. 8 crosscut handsaw
1 framing square
1 6-ft folding ruler
1 chalk line
1 16-oz claw hammer

ESTIMATED MANHOURS

1.5 hours per 16-ft run of bridging

PROCEDURE

Step 1　Measure in 6 to 8 ft from a reference point at two places and snap a chalk line.

Step 2　Install a spacer board, previously marked, 15 in. from the chalk line.

Step 3　Cut one piece of bridging for each common space.

Step 4　Measure and cut a piece of bridging for each nonstandard space between joists. Mark the piece to identify its future position.

Step 5　Nail the bridging in place, alternating pieces on opposite sides of the chalk line.

9

SHEATHING

BASIC TERMS

C-C grade plywood knot holes 1 in., occasionally larger, on both sides of plywood, sanded or unsanded, for rough construction and base for tile and linoleum.

Exterior-type plywood 100 percent waterproof glued plys; inner plys exceed the grade of interior plywoods.

Fiberboard board made from farm by-products and saturated with tar-base materials.

Hip rafter on a roof extending from a corner up and back from two directions simultaneously.

Overhang that portion of the roof which extends beyond the outside wall line.

Ridge uppermost horizontal member of a roof framework.

303 Siding C plugged or better-grade exterior plywood for finished purposes; 3/8 in., 1/2 in., and 5/8 in. may be grooved.

Standard-grade plywood C-D grade sheathing, where interior plys are D grade, used for subflooring wall and roof decking.

T-1-11-type plywood exterior-type panel with shiplap edges and parallel grooves 1/4 in. deep, 3/8 in. wide, 2, 4, 6, or 8 in. OC.

Valley joint roof line where two A-frame roofs join; type of rafter.

The enclosing of new work—framing, flooring, and roofing—requires the use of various materials. These materials made from wood and fiber products, must be selected for the function they are to perform. In every application, the work and material are called *sheathing*.

SCOPE OF THE WORK

Considerable time and money will be spent on installing sheathing. The closing-in process requires the use of sheathing-grade plywood, fiberboard, and stock lumber. Prior to the days when sheathing-grade plywood was available, stock lumber (1 x 6 or 1 x 8, often tongue and groove) was installed. It is still available today and frequently used for remodeling and repairing tasks.

Sheathing is installed in three major areas on a new addition (Figure 9-1). They are subfloor over joists and the exterior walls on the studs and on the rafters. Each application requires its own special considerations. Subflooring of the plywood variety may be installed in one or two layers. Plywood sheathing may be used on walls by installing the sheets vertically or horizontally. Plywood sheathing in combination with fiberboard is frequently installed on exterior walls. Sheathing-grade plywood of proper thickness and type is used to close in a roof.

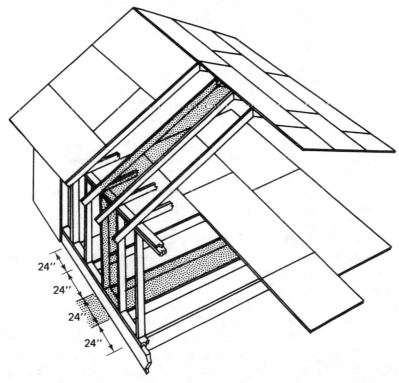

Figure 9-1 Sheathing

There are three major areas in construction where sheathing is used. This section presents specific information for each area.

SHEATHING SUBFLOORS

Plywood is an excellent product to use when installing subflooring over floor joists as shown in Figure 9-2. Some of its advantages are its inherent squareness, its uform thickness, and its rigidity. In addition,

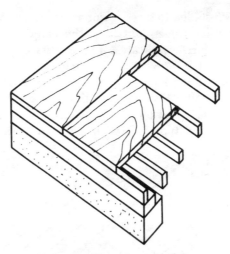

Figure 9-2 Sheathing a floor

it is sold in thicknesses from 1/4 to 1-1/4 in. Two types of installation may be used, each with its own characteristics: the single-layer method and the double-layer method.

To install the proper type, it is important to know which plywood grades and sizes are used for each method.

Grades and Sizes of Plywood

The grade of plywood most commonly used for subfloor installation is the standard C-D interior-grade plywood or grade C-C exterior-grade plywood. If the plywood will be installed in moist or high-humidity locations, the exterior type is far superior because of its glue.

The plywood usually used for the first layer of subflooring is either 5/8, 3/4, or 7/8 in. thick. As a rule it does not need to be smooth-sanded.

If a single layer of plywood is to be installed on floor joists and carpeting is to be laid over it as shown in Figure 9-3, one thickness of plywood 1-1/4 in. thick may be used. A type available on the market is labeled "underlayment C-C grade plugged, 2-4-1." This type of plywood also makes an excellent base for parquet floors and all types of resilient floors where the total floor thickness equals 1-1/2 to 2 in. above the joist.

If a double layer of plywood is to be installed over joists as shown in Figure 9-4, the first layer should be the standard C-C or C-D grade. The second layer should be C-D or C-C, where the C side is both plugged and sanded. Its thickness should be selected so that the flooring tile and plywood thickness combine to identify a finished floor height.

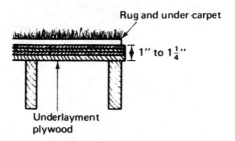

Figure 9-3 Single-Layer Subfloor Under Carpet

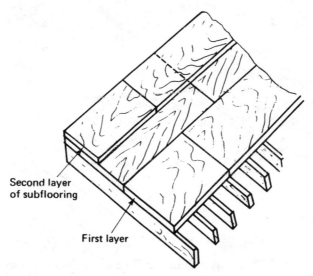

Figure 9-4 Double-Layer Subflooring

It is important that a sanded surface of plywood be made available for gluing tile and other resilient types of flooring. This second layer of plywood should be installed just prior to installing the tile to prevent it from becoming contaminated and marred.

Squaring

Each standard sheet of plywood is 4 ft x 8 ft and the sides are square. Because of this feature the plywood should be used as an aid to squaring any slight irregularities in the subfloor joist framing. Figure 9-5 shows two methods that may be used in the squaring process. Figure 9-5A shows how the box beam and end beams in a joist foundation are formed to the edges of the plywood. The second and remainder of the sheets are aligned in the same manner. The second method (Figure

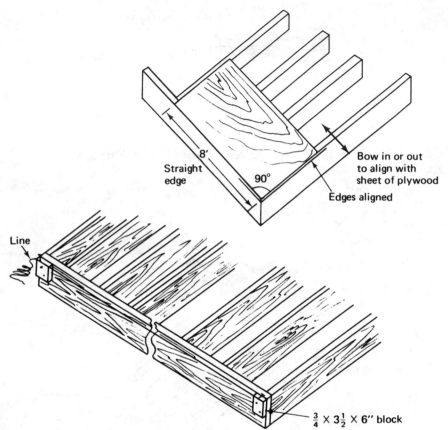

Figure 9-5 Squaring a Subfloor Foundation with Plywood

9-5B) shows that a taut line is placed 3/4 in. off the outer edge of the box beam (or along open-beam construction). All sheets of plywood are aligned 3/4 in. back from the line by use of a 3/4-in. block of wood.

Breaking Joint

The next characteristic in laying sheathing subflooring is that of *breaking joint*. This means that the end of a sheet on adjacent rows will not fall on the same joist. Plywood manufacturers recommend that each sheet of plywood extend over two joist spaces (spans). This means a minimum length of 32 in. where joists are spaced 16 in. OC and a minimum length of 48 in. where joists are spaced 24 in. OC (Figure 9-6).

If a second layer of plywood is installed, it should break joint differently than the ends of the subfloor and should also break joint along the long side (row side) of each row.

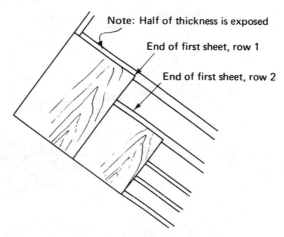

Figure 9-6 Breaking Joint

Nailing Technique

Each sheet that is installed should be nailed with the size of nail selected from Table 9-1 based upon the thickness of plywood. Nailing along the outer edges of each sheet should be held to 6-in. intervals, whereas nailing across the sheet may be expanded to 10-in. intervals.

TABLE 9-1: Nailing Plywood

Plywood Thickness (in.)	Nail Length	
	Minimum	Maximum
1/4	3/4 in., 3d	6d
3/8	3d, 4d	6d
1/2	4d	6d, 8d
5/8	6d	8d
3/4	6d	8d

Edge Support for Plywood

The ends of a sheet (4 ft width) are nailed on a joist in every case; however, the 8-ft edge is not. Therefore, 2-in. blocking must be installed below the edge of the sheet so that the edges of adjoining sheets may be nailed at 6-in. intervals (Figure 9-7).

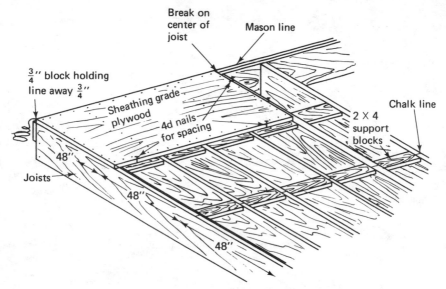

Figure 9-7 Blocking for Edge Nailing of Plywood

Spacing of Plywood Sheets to Allow for Swelling

In humid climates or where moisture may come in contact with subfloors, each sheet of plywood must be allowed to expand. The simplest method for allowing swelling is to use 4d nails as spacers between sheets of plywood. Use 6d nails between sheets if heavy swelling is anticipated. With the application of these installation characteristics you should have no adverse effects when the job is completed. The floor should be sound, rigid, and should never develop squeaks or creaks. But, what about sheathing exterior walls?

SHEATHING WALLS

Plywood by itself is an excellent material to use when covering exterior wall framing. However, a fiberboard impregnated with tar-base materials is often used in combination with plywood. For repairs and remodeling either or both products mentioned above may be used; but do not forget the versatility of 1 x 6 stock lumber sheathing.

Grade and Type of Plywood

Exterior-type C-C or Structural 1 types of plywood are recommended for exterior-wall application. Both of these types of plywood are available in unsanded and touch-sanded finishes.

Fiberboard

Fiberboard made for exterior sheathing combines in one panel a wall-covering material, an insulation, and a moisture-retarding barrier. These sheets are available in 1/2-, 5/8-, and 3/4-in. thicknesses and in standard 4 x 8 ft and 4 ft x 10 ft sizes as well as 2 ft x 8 ft tongue-and-groove style. The board selected to be used along with plywood sheathing must be equal in thickness to the plywood. Figure 9-8 shows how the board looks when installed. This figure shows a vertical installation; however, the panels may be installed horizontally.

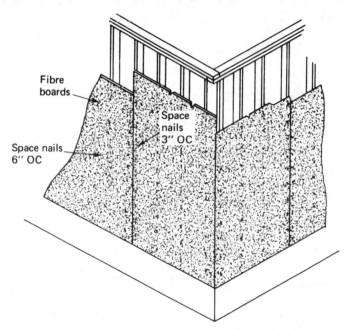

Figure 9-8 Vertically Installed Fiberboard

The holding power of fiberboard is not very good, so its use must be selectively applied. If a siding such as asbestos shingles (pieces 12 in. x 24 in.) or wood shingles are to be used, fiberboard should not be used. On the other hand, if brick, lapsiding, or plywood paneling such as T-1-11 (grooved panels) or 303 special (grooved) is used, the fiber panel is an excellent choice. Nailing to studs is possible through the outer plywood or siding surface, through the fiberboard, and into the studs.

Outside Corners

Outside corners are usually sheathed with plywood panels installed vertically. This type of installation, also shown in Figure 9-7, adds strength and rigidity to the framing structure.

Vertical Installations

If plywood and/or plywood and fiberboard is to be used to sheath exterior walls, the vertical method provides an excellent opportunity to completely enclose the wall without having any splices. The panels should be selected for a length that will reach from foundation to overhang. The break along the edge will center on a stud and the ends will be even with sill and plate. Window and door openings may be cut before or after the sheets are installed.

SHEATHING ROOFS

Sheathing a roof with plywood is an excellent, time-saving method. It is unlike wall sheathing because fiberboard cannot be used. Since it has its own characteristics, let's identify them by starting with grades and types.

Grades and Types of Plywood Used for Roof Sheathing

Roof sheathing should be grade Standard, Structural I or II, or C-C-Exterior. All grades must be classed as exterior. Usually 1/2-in. plywood is used for roofing; however, in some localities building codes permit 3/8-in.-thick plywood. Whether using 3/8- or 1/2-in.-thick plywood, use plyclips between roof-rafter spacing and sheets of plywood as shown in Figure 9-9. These clips are H-shaped and support both sheets in the space between roof rafters.

Breaking Joints

When installing sheets of plywood, be sure to have the ends of the sheets break on different rafters on successive rows. Space the ends of sheets 1/16 in. as a rule, in high-humidity areas 1/8 in., before nailing.

Trimming Overhangs

Trimming overhangs on A-framed roofs is frequently done after the plywood is installed. Usually a chalk line is snapped and a power saw is used to trim excess plywood. This procedure makes for a neat, straight roof line.

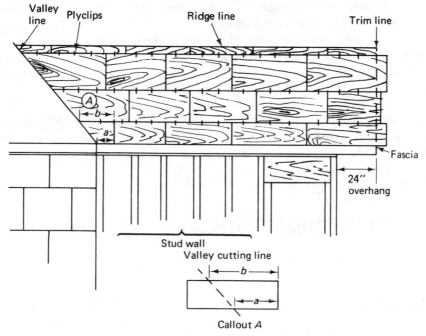

Figure 9-9 Plyclips Between Sheets of Plywood

The procedure is usually not applicable to overhangs that are paralleled to the ground because this is the starting point for sheathing.

Valley, Hip, and Ridge

Plywood pieces that are to be installed in valleys are usually cut on the ground. A person on the roof measures its size, and a person on the ground cuts it. It is not practical or safe to cut this piece on the roof (Figure 9-9).

On the other hand, plywood may easily be installed over a hip and the overhang may be trimmed with a power saw. If a handsaw is to be used, the pieces will need to be measured, then precut on the ground.

The last row of plywood ends at the ridge. These pieces are usually measured, then cut on the ground. This practice is the safest method. Follow the mailing suggestions given in Table 9-1.

ROUTINES

Next are three routines detailed for this type of work. Each one provides all the required data about that phase of sheathing.

To complete your material requirement you will need to calculate the total number of square feet for each job. Refer to Chapter 3 if you need a refresher on calculating this type of data. You can expect to use each routine at a different time in your job plan. Each application will probably require a different type and thickness of plywood, so no problems can be expected when ordering the material.

This is one work area where a lot of visible progress is seen for a little time expended. Do not forget that you will probably need a helper for this job. Also remember to space your sheets to allow for swelling, block the edges to prevent squeaking, and nail securely.

BS1: SHEATHING SUBFLOORS

RESOURCES

Materials:

_____ sheets 4 ft x 8 ft x _____, _____, and _____
    no.   thickness   grade   type

of plywood. No. = floor _____ × _____ ÷ 32 sq ft which
                          length (ft)   width (ft)

= the total number of sheets. Add 1 to 2 sheets to allow for waste.

_____ lb 8d common nails. 1/2 lb per sheet is required.

_____ ft 2 × 4 blocking. The number of lineal feet for one row ×
    L

the number of rows equals the total lineal feet. Add 10 percent for waste.

_____ lb 8d common nails for toe nailing. 2 × 4 blocking = 1 lb
    no.

nails per 6 lineal ft.

Floor plans

Specifications: data as to type, grade, and thickness of plywood

Tools:

1 pair sawhorses
1 chalk line
1 framing square
1 crosscut handsaw

1 portable power saw 6 to 7-1/4 in. (recommended) and 50-ft extension
cord
2 16-oz claw hammers
1 100-ft mason line
blocks of 3-3/4 in x 2 in. 6 to 8 in. long for standoff line
1 6-ft folding ruler

ESTIMATED MANHOURS

For two people   3 hours
based upon an 180-sq-ft area; includes 2 × 4 blocking (approximately
1 minute per square foot)

PROCEDURE

Step 1 Snap chalk lines at 4-ft intervals from the leading edge and
perpendicular to the joist run.

Step 2 Cut and install 2 x 4 blocking by centering the block on the
chalk line with the 3-1/2-in. surface flat; toe-nail as per Figure 9-7.

Step 3 If using the standoff string method, do steps a and b. If not,
skip to step 4.
a. Tack-nail a piece of 3/4 in. x 2 in. x 6 in. stock below the top of the
   joist edge on each end of the joist foundation (Figure 9-5B).
b. String a line (tautly) between the two nailed pieces.

Step 4 Position the first sheet of plywood by breaking its end on a
joist and aligning it with the joist end or 3/4 in. back from the string;
nail with 8d common nails.

Step 5 Insert two 4d nails into the joist (partially driven) at the end
of the first sheet. Butt the next sheet against these nails and align with
the leading edge of joists or 3/4 in. back from the string line and nail.

Step 6 Repeat step 5 until the row of sheathing is complete except for
the last sheet.

Step 7 Precut the last sheet of plywood for length or install and cut
as required (optional).

Step 8 Precut a sheet of plywood by one joist spacing to begin the
second row, or install the sheet and trim after. (*Exception:* Where less
than one-half the width of a sheet extends past the ends of the joist,
trim the excess after the row is completed.) Space the edges of the sheets
with 4d nails.

Step 9   Repeat steps 5 through 8 until all but the last row is complete.

Step 10   Rip the plywood for the last row prior to installing it in most cases.

BS2: SHEATHING EXTERIOR WALLS

RESOURCES

Materials:
Specifications: for grade, size, and thickness and type(s) of material

(Plywood only) _____ sheets 4 ft x 8 ft x _____, _____,
                no.   thickness   grade

and _____ of plywood. No. = total square feet of wall (length
      type

in feet of perimeter × height in feet) divided by 32 (number of square feet per sheet) plus 10 percent of the number of sheets for waste.

(Plywood & Fiberboards)

Plywood _____ sheets 4 ft x 8 ft x _____, _____, and
             no.   thickness   grade

_____. No. = one sheet for each outside corner of the exterior
    type

walls.

Fiberboard _____ sheets 4 ft x 8 ft x _____, _____, and
              no.   thickness   grade

_____. No. = total square feet of exterior wall surface divided
    type

by 32 sq ft per panel less the number of outside corner plywood sheets plus 10 percent of the number of sheets for waste.

_____ lb 8d common nails. 1/2 lb per sheet is required.
    no.

Tools:
1 8-ft to 12-ft ladder
1 pair sawhorses
1 pair 2 x 10 or 2 x 12 scaffold planks
1 24-in. level
2 16-oz hammers

1 portable power saw (recommended; or 1 brace, 1 3/4-in. bit, and
   1 keyhole saw)
1 6-ft folding ruler
1 no. 8 crosscut handsaw

ESTIMATED MANHOURS

For two persons   2.2 hours
average 1 minute per square foot; based upon four-sheet installation
   with one door and one window cutout

PROCEDURE A: VERTICAL INSTALLATION

Step 1   Plan the installation starting point on the wall. (Measure in
from the outside corner to locate 48-in. stud placement.)

Step 2   Insert two 16d common nails between the sill and the cement.

Step 3   Precut the panel to the exact height needed (if required).

Step 4   Position on nails, align with outer corner, and center on 48-in.
2 x 4 stud; nail in place at 6-in. intervals, 10-in. intervals in the center of
the sheet.

Step 5   Repeat steps 2 through 4 for each sheet.

Step 6
a. Cut out window and door openings by driving a nail from the inside
   of the wall at the four corners of the intended opening.
b. Connect nails with a pencil line and level. Cut out with the power saw.
   (See alternative to tool list if a power saw is not available.)

PROCEDURE B: HORIZONTAL INSTALLATION

Step 1   Insert two 16d common nails between the sill and the cement.

Step 2   Measure from the corner 8 ft to determine 8-ft stud placement.

Step 3   Position the sheet horizontally with the corner and end even
and the other end centered on a stud; nail with 8d common nails.

Step 4   Remove two 16d common nails from the sill/concrete and reinstall for a second sheet.

Step 5   Install two 4d common nails into the 2 x 4 at the end of the
first sheet.

Step 6   Position the second sheet against all four nails and nail in place.

Step 7 Repeat steps 4, 5, and 6 for all remaining sheets in the row except the last sheet.

Step 8 Trim the last sheet to the needed length before positioning and nailing.

Step 9 Cut any partial window and door openings before installing the second row of sheathing.

Step 10 Trim a sheet of plywood by one stud spacing and install from the corner for the second row.

Step 11 Place 4d spacer nails between the sheet edges and nail the sheet in place.

Step 12 Remove the 4d nails and reinsert them for the next sheet; also install two 4d nails at the end of the first sheet.

Step 13 Install the second and remaining sheets in a similar manner except for the last sheet.

Step 14 Trim the last sheet before nailing it in place.

Step 15 Cut out the remainder of the window and door openings.

BS3: SHEATHING A ROOF

RESOURCES

Materials:
Specifications: estimate of plywood for roofing
Plans: roof layout plan
_____ sheets 4 ft x 8 ft x _____, _____, and _____
    no.   thickness   grade   type

of plywood. No. sheets equals total square feet of roof (both slopes) divided by 32 plus 10 percent of the number of sheets for waste.

_____ lb plyclips. 1 lb equals enough clips for an average 320 sq ft
    no.

of roofing.

_____ lb of 6d common nails. 1 lb equals enough nails for an average two sheets.

Tools:
2 16-oz hammers

1 no. 8 crosscut handsaw
1 portable power saw (recommended)
1 pair sawhorses
1 12-ft extension ladder
1 chalk line
1 6-ft folding ruler
1 framing square
1 pair roof jacks (optional, if required)
1 pair 2 x 6s 12 ft long for roof scaffold planks (optional, if required)

ESTIMATED MANHOURS

For two persons   4.24 hours for 500 sq ft
average 2 minutes per square foot

PROCEDURE

Step 1 Determine the position of the first sheet of plywood with the aid of your plan.

Step 2 Install the first sheet with the lower edge flush with either the rafter or the fascia *and* properly center on a rafter with sufficient overhang on the gable end.

Step 3 Nail the sheet with 6d common nails at 6-in. intervals on ends and at 10-in. intervals through the center of the sheet.

Step 4 Partially drive two 4d common nails into the rafter alongside the end of the first sheet.

Step 5 Align the second sheet with the lower roof edge and butt against 4d nails. Nail in place.

Step 6 Complete the row by repeating steps 4 and 5.

Step 7 Trim a sheet of plywood by one rafter spacing.

Step 8 Insert a plyclip between each rafter on the upper edge of the first row of plywood sheathing.

Step 9 Repeat steps 4 through 8 until all the rows are installed *except* the last.

Step 10 Precut by ripping the final row of plywood on the ground before installing on the roof.

Step 11 Measure the gable overhang at the lower roof edge and ridge. Snap a chalk line trim with the portable power saw.

ALTERNATIVE PROCEDURE

Alternative step 1   Cut valley pieces of plywood by measuring its size at both the lower and upper edges of the previously installed sheet (Figure 9-9A).

Alternative step 2   Trim the hip rafter sheathing excess by snapping a chalk line and cutting with the power saw, or precut on the ground, then install.

10

ROOF FRAMING

BASIC TERMS

Bird's mouth cut made in a rafter to make the rafter seat on the wall's plate.

Collar beams nominal 1- or 2-in.-thick members connecting opposite roof rafters. They serve to stiffen the roof structure.

Common rafter rafter whose ends fasten to a ridge and plate and may or may not extend past the plate to make an overhang.

Fly rafter end rafters of the gable overhang supported by roof sheathing and lookouts.

Gable end triangular vertical end of a building formed by the eaves and ridge of a sloped roof.

Heel cut vertical part of a bird's-mouth cut.

Overhang amount or distance that a rafter projects past the outer wall surface.

Pitch value assigned to the slope of a roof determined by dividing 24 into the rise in inches to obtain a fractional quantity (e.g., 8-in. rise = 8/24 = 0.333, which translates to 1/3 pitch).

Rafter rise number of inches that a roof's rafters will rise per foot of run.

Rafter run equal to one-half the building's span (width). The rafter run, in combination with the rafter rise, determines the rafter length.

Ridge highest horizontal member on a roof to which rafters are nailed.

Ridge cut cut on a rafter that allows the rafter to fit to the ridge member.

Span total width of a building on a line with the rafter run.

Valley exterior corner in a roof design.

Valley jack rafter rafter whose ends fasten from a valley to either a ridge or a plate.

Roof construction usually requires a considerable amount of construction knowledge on the part of the builder. Of all aspects of the new construction effort, this area of work contains the most difficult laying-out tasks. The primary cause is laid to the numerous applications of bevel and combined miter cuts.

This chapter provides details in simplified step-by-step procedures which, when understood and followed, will result in successful framing of either a gable A-framed roof or a lean-to roof.

SCOPE OF THE WORK

Much of the work involved in planning, laying out, and cutting rafters, ridges, and gable-end studs should be done on the ground.

Bracing, first temporary, then permanent, must be installed to keep the roof line plumb, provide rigidity, and reduce the outward stress placed upon the bearing walls.

If a gable roof ties into an existing roof at a 90-degree angle, a special rafter, called a *valley jack rafter*, must be cut and installed over the old roof area. As its name implies, the rafter will be instrumental in framing a valley between the old and the new roof. These rafters are usually difficult to lay out and cut, but by using Routine BRF4 after studying this chapter, you should have no trouble.

With the framing complete, you will need to use Routine BS3, Chapter 9, to aid you in sheathing the roof.

Except for the tying-in methods, the parts of a roof and the method used to lay them out and cut the various members are essentially the same for the gabled roof and the lean-to roof. During this study of roofs, their rafters, tying-in methods, and bracing, several principles are discussed in detail. These principles are provided for the purpose of familiarizing you with some of the technical aspects of roof construction. Because of the manner in which the routines are developed (following this chapter), you will be using the data, but it will be disguised in steps of the routines.

RUN AND RISE

Before discussing the various parts of a rafter, let us apply and understand two significant terms in roof construction. These are the rafter *run* and rafter *rise*. As you can see in Figure 10-1, the rafter run is equal to one-half the overall span of the building. If your building has a 20-ft-wide span, the run is 10 ft long, or one-half the span. This is a fixed rule.

The *rise* is the vertical height of the roof measured from the plate to the top of the ridge. The rise may be calculated as follows, but, more important, the rise represents the specific number of inches that a roof's slope will have for each 12 in. of rafter run.

Following is the *formula for calculating a roof's total rise:* Multiply the roof's rise in inches × the number of feet and fractional feet per rafter run. For example, a 10-ft rafter run with a 5-in. rise is a total of 50 in. of rise.

Figure 10-1 shows a symbol near the roof line that is similar in shape to an inverted right triangle. This symbol provides you instantly with details about your roof's run and rise. The top number (12) informs you that the rafter is *common,* and the other (side) number indicates the

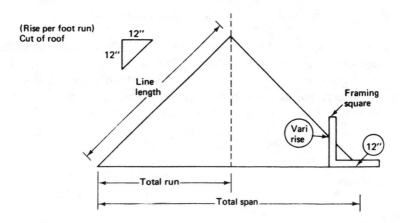

Figure 10-1 Rafter Run and Rise

rise in inches per foot of run. (Note: On hip or valley rafters, not discussed in this book, the 12 would be replaced with a 17.)

What use can you make of these facts? You can use them to obtain the total rise, and they will tell you how to position your framing square to make a ridge cutting line, bird's-mouth cutting lines, and overhang cutting lines.

Figure 10-2 shows a framing square with numerous rises per foot of

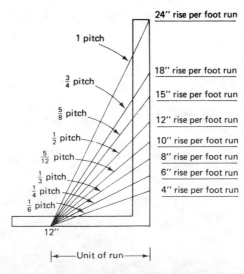

Figure 10-2 Framing Square with Rafter Rises

run. As soon as practical, try to use your square on a scrap piece of 2 x 6 to get the feel of this part of the task. As an example, try laying the square in place for a 4-, 5-, and 8-in. rise per foot of run. Draw a line along the outer edge of the square's tongue edge. Follow this activity by standing the scrap piece so that when a level is held *plumb* with the mark drawn, you can conceive the impression of a sloping rafter (Figure 10-3). Note the gentle slope of the 4- and 5-in. rise and the steepness of the 8-in. slope.

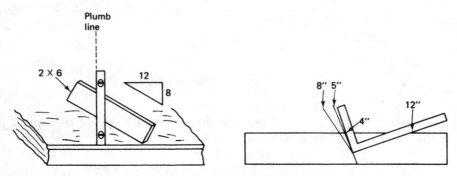

Figure 10-3 Visualizing a Rafter Slope

PARTS OF COMMON RAFTERS

Ridge Cut

The *ridge cut,* shown in Figure 10-4, is precisely made where its cut surface is perpendicular to the plate's run. This accuracy is essential. You

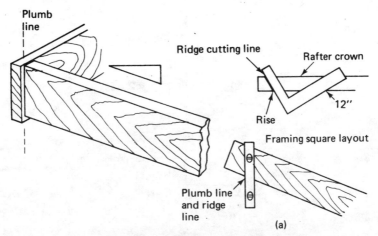

Figure 10-4 Ridge Cut on Common Rafter

can make this layout with your framing square by placing the square with the 12-in. mark on the blade and the rise (in inches) on the tongue on the crown side of the rafter. However, the method explained later will relieve the need for using the square to make this mark. Instead, a level will be used in plumb fashion as shown in Figure 10-4A.

Bird's-Mouth Cut

The second cut needed on a common rafter is the *bird's mouth*. This cutout in a rafter looks, as shown in Figure 10-5, and has two surfaces: the heel cut and the flat cut. The *heel cut* is the exact angle of the rise per foot of run or the ridge cut. The *flat cut* is a horizontal cut which is needed so that the surface will rest evenly on the wall's plate. The corner between the heel and flat cut is always 90 degrees. This layout can be made with a framing square or with a level, as we will see later. To make the layout with a square, make a line representing the ridge cut on a scrap piece of 2 x 6. Next, lay the square flat upon the 2 x 6 aligning the blade with the line already drawn and the tongue positioned as shown in Figure 10-6. Slide the square up or down along the line until

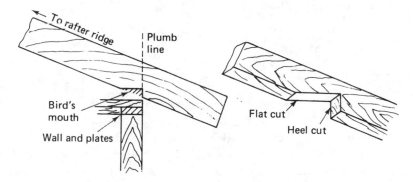

Figure 10-5 Bird's Mouth Cutout

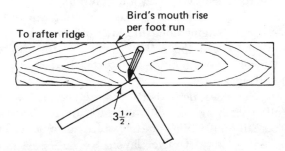

Figure 10-6 Laying Out a Bird's Mouth with a Square

the tongue marking of 3-1/2 in. is flush with the edge of the 2 x 6. Draw a line from the edge to the intersecting line. The layout is complete.

Overhang Cut

The third cut needed on a common rafter is the *overhang cut*. The perpendicular surface is laid out opposite to the ridge layout where the square's mark of 12 in. and the rise are located along the bottom edge of the rafter. There may be a requirement to lay out and cut a *flat cut* on the overhang. This condition is explained in Figure 10-7A where, for example, a 1 x 6 fascia board is to be installed. Note that of the total 5-1/2 in. board, there is only approximately 4 in. available that can be against the rafter. Since the 2 x 6 rafter's plumb-cut surface is approximately 7 in. long, some must be cut off the bottom of the rafter end so that the soffit may be installed.

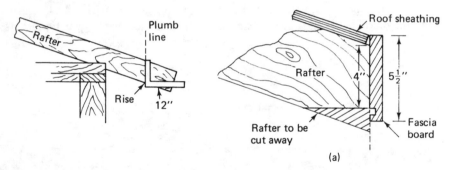

Figure 10-7 Overhang Layout

One more aspect of the rafter end needs to be described, how to measure and mark the amount of overhang. Customarily there are three dimensions of overhang: (1) none (Figure 10-8), (2) 12-in. overhang, and (3) 24-in. overhang. Where no overhang is used, part of the bird's-mouth layout forms the rafters' outer end. For the applications containing an overhang, the distance can easily be determined if the bird's-mouth layout is used as a reference point. Use a framing square as shown in Figure 10-8 and lay its tongue alongside the bird's-mouth heel mark. With the blade pointing toward the rafter end, mark the desired overhang projection on the rafter. Once marked, the standard layout can be made for rafter-end cutting.

Frequently, there are methods that translate principles into working techniques. Following is an accurate and fail-safe method for laying out your rafters. You can do it while standing on the ground and without having an in-depth knowledge of rafter layout.

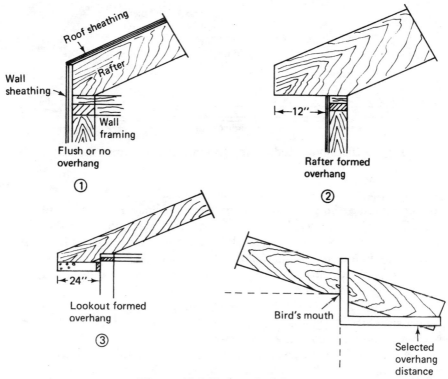

Figure 10-8 Rafter Overhang

COMMON RAFTER LAYOUT: SHOP METHOD

Review and study Figure 10-9 until you have located the following data:

1. The building span line.
2. The rafter run line.
3. The rafter rise line.
4. The rafter slope line.
5. The 2 x 5 block that represents the wall's plate.

Objective: Simulate a rafter's position by using the wall (or driveway) as a reference area and formally lay out and fit a rafter to the layout.

Snap a plate-reference chalk line from 12 to 18 in. above the sill of the wall. This will allow you to project the portion of rafter needed for overhang beyond the end of the wall and still be above ground level with its end.

You must calculate the total rise of the rafter by using the formula

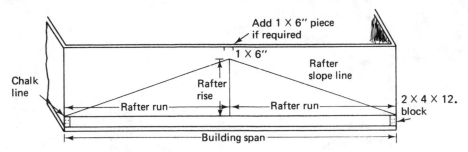

Figure 10-9 Shop Method for Laying Out a Common Rafter

provided previously (rise × rafter run = total rise in inches). The line for the rise must be perpendicular to the base line so that a plumb bob can effectively be used. Measure the rafter run along the base line. Suspend the plumb bob above the mark on the base line and mark the upper end of the plumb bob's line. Snap a line connecting both points. If the total rise is greater than the height of the wall, temporarily nail a 1 x 6 or 2 x 6 approximately centered where the vertical line is to be snapped. Snap the line on the board. Measure *up* from the base line and mark the total rise. Then snap the slope (hypotenuse) lines between the total rise mark and the outer corners of the base line.

Next, nail a scrap piece of 2 x 4 x 12 in. alongside the corner stud with the block's *top* flush with the base chalk line. Drive a 16d common nail partially into the 1 x 6 cr 2 x 6 (if used) at the intersect point of the vertical span (center) line and the slope line.

Lay a rafter on the nail and 2 x 4 block and, after ensuring that the ridge end is fully beyond the center line, tack the rafter in place with an 8d common nail.

Marking a Bird's Mouth

Use a 24-in. level and a pencil to make a bird's-mouth layout as shown in Figure 10-10 (which is an extract from Figure 10-9). Remove the rafter and cut out the bird's mouth.

With your ruler measure the depth of the bird's-mouth *heel*. (Refer again to Figure 10-10.) Lower the nail driven into the intersect point of the slope and vertical lines by the amount measured. This step is necessary to maintain the exact rise per foot of run called for.

Reinstall the rafter on the layout with its bird's-mouth cutout seated tightly against the 2 x 4 block. Tack the rafter in place.

Laying Out a Ridge Cut

Use your level again; lay out the ridge cut, as shown in Figure 10-11 (which is an extension of Figure 10-9). Next, subtract one-half the thick-

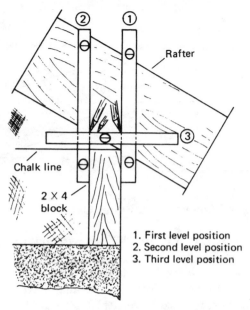

1. First level position
2. Second level position
3. Third level position

Figure 10-10 Laying Out a Bird's Mouth with a Level

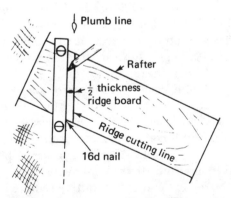

Figure 10-11 Laying Out a Ridge Cut with a Level

ness of the ridge board by measuring in a distance from the vertical line just drawn. Draw another vertical line even with the mark just made and label it "cutting line." While the rafter is in position, the overhang should be laid out, after which all cuts to the rafter can be made.

Laying Out an Overhang

Only one example is provided of the laying out of an overhang. You must use the principle to fit your specific requirements. Figure 10-12

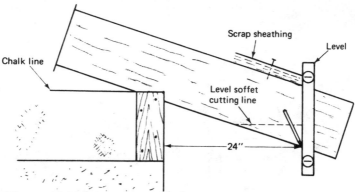

Figure 10-12 Laying Out a Rafter Overhang with a Level

(which is an extension of Figure 10-9) shows how the principle is employed.

Assume that a 24-in. overhang is being laid out. Measure on a horizontal plane (from the 2 x 4 block) 24 in. and make a mark. Use your level in plumb fashion to make a vertical line.

Next, review your plans and determine the width of the fascia board so that you can lay out the amount of rafter-end lower edge that needs to be trimmed. Also note how much is needed/used to retain the soffit if a box cornice is installed. The plan should show you this detail, or you may calculate it as follows. Subtract the lip and the soffit's thickness from the total width of the fascia board. Record your remainder.

Lay a piece of roof sheathing material (a scrap piece) on the rafter's top edge, and with your ruler measure down from the plywood's top edge a distance equal to that recorded. Mark along the vertical line previously drawn. Draw a level line from this mark *back* toward the bird's mouth. Your rafter layout is complete. Cut out the rafter and another one. Try them for fit on the opposite sides with a scrap piece of stock that represents the ridge, inserted. If they fit (which they should), use the rafter as a pattern to lay out and cut all rafters required. *Caution:* Be sure to keep the crown *up* on all rafters.

MARKING PLATES AND RIDGE BOARDS FOR RAFTERS

Maximum strength in building construction is achieved when the joists and rafters are installed directly over the wall's studs. Therefore, start your layout on the same corner that you did for joist and stud layout. Space your rafters 16 or 24 in. OC as you did the studs. Make an X where the rafter will actually rest. Mark a ridge board according to the plate layout and, *if needed,* extend the ridge over the wall to account for a gable-end overhang. Trim the inside end of the ridge board so that its end centers on a rafter. This allows the second ridge piece to be nailed to the same rafter.

RAISING A ROOF

The rafters are installed in a precise manner. The end or outer rafter (not the gable overhang) is customarily one of two rafters installed first. The second rafter is installed 6, 8, or 10 ft farther along the ridge. Both rafters are installed on the same side of the ridge (Figure 10-13).

With the help of a second person, raise the ridge and rafters until the bird's mouths are seated properly against the plate of the wall. Nail the rafters to the plate, and then tack two braces to the rafters (near the ridge) to hold the ridge up and free the man holding the ridge.

Install rafters in sets of two on one side of the ridge, then two on the other side. This ensures that your ridge line will be straight.

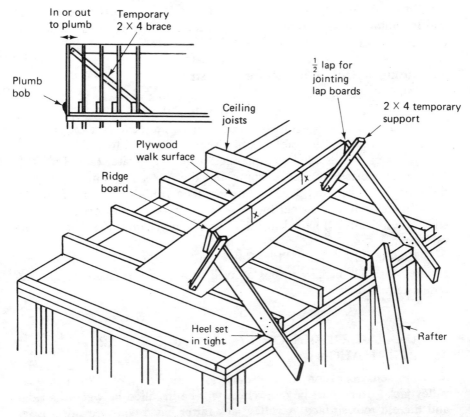

Figure 10-13 Raising a Roof

When the first 8, 10, or 12 ft of roof is raised, an alignment must be performed. Use your plumb bob to plumb the end rafter at the ridge even with the outer edge of the plate. When in line temporarily, nail a brace from the ridge to the ceiling joists at about a 45-degree angle.

Bracing the Roof

Continue the installation of rafters and ridge until all are installed. When complete, a permanent brace must be installed at each gable end of a gable roof. Routine BRF3, Roof Bracing and Collar Beams, should be used as an aid to cutting and installing these braces (Figure 10-14).

Installing Gable-End Studs

Following this activity, the gable-end studs can be installed. These studs should be aligned over the wall stud so that maximum strength is achieved. A special cut is made on the upper end of each stud. Part of the stud is flush with the outer edge of the rafter. Figure 10-15 shows a gable-end stud with the cut for the rafter completed. Routine BRF5, Laying Out and Installing Gable-End Studs, outlines the basic procedures needed for the task.

Rake or Gable-End Rafter Construction

The rake section, shown as one example in Figure 10-16, is the extension of a gable roof beyond the end wall of the house. This detail may be classed as: (1) a close rake with little or no projection, or (2) a boxed or open extension of the gable roof, varying from 6 in. to 2 ft or more. A *fly rafter* must be installed on a planned overhang greater than 8 in. This rafter is supported by roof lookouts and the roof's sheathing. Note that the lookouts, spaced on 24-in. centers, extend from the second rafter on the roof through notches cut into the end rafters and out over the gable end. The fly rafter is nailed to the ends of the lookout after making the ridge cut.

Routine BRF7, Installing Gable-End Overhang Rafters, provides detailed installation instructions as well as a guide to calculate the materials required.

CHARACTERISTICS OF VALLEY JACK RAFTER INSTALLATION

On some occasions a roof must join another roof. When it does, valley jack rafters must be cut, which will be installed between the ridge and the old roof surface. A valley jack rafter has a common rafter ridge cut. The other end of the rafter has a special compound miter cut. Since

the valley jack rafter valley cut will lay on the old roof's sheathing or a
1 x 8 or 2 x 8 piece of stock nailed along the valley line, its cut is different from customary valley cuts.

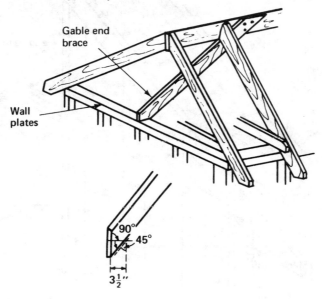

(a) 2 × 4 brace

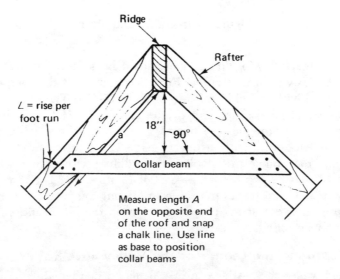

(b) Collar beams

Figure 10-14 Bracing a Roof

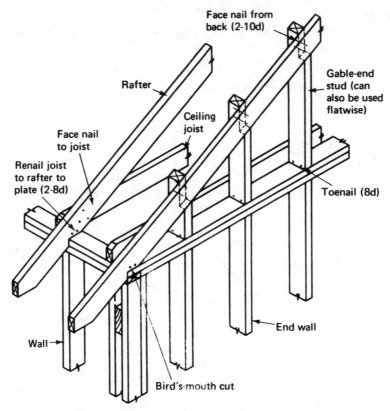

Face nail from
back (2-10d)

Rafter

Gable-end
stud (can
also be used
flatwise)

Ceiling
joist

Face nail
to joist

Renail joist
to rafter to
plate (2-8d)

Toenail (8d)

End wall

Wall

Bird's-mouth cut

Figure 10-15 Gable-End Stud

From the following image you should be able to lay out a valley rafter cut with very little trouble. First, recognize that the long cut must be horizontal with the old roof line (Figure 10-17). The cut is made by using the standard rise per foot of run and 12 in. However, instead of marking the rise (tongue on square), mark the run along the blade on the square. The other cutting line is made to allow the rafter to set on the sloping roof. Since this sloping angle differs with each pitch, a simplified table has been provided as part of Routine BRF4. Use your framing square as suggested in the table to lay out this angle cut. Place the compound miter cut at the proper position on the rafter and your rafter is ready for installing.

Measuring the Length of the Valley Jack Rafter

Each valley jack rafter can be calculated mathematically; however, the shop method, although requiring a bit more physical labor, will re-

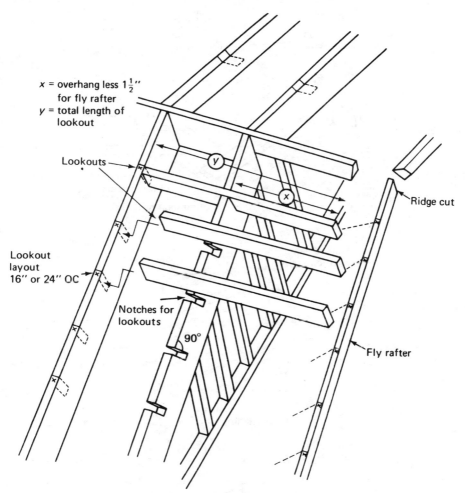

Figure 10-16 Fly Rafter Installation

sult in a clear understanding of the length of each rafter. This means going up on the roof and measuring each rafter.

The first step requires that you lay out each rafter on a center position on the ridge that extends to the old roof. The second step requires you to snap a chalk line identifying the valley line. It will extend from the top and end of the ridge board, where it intersects the old roof sheathing, to the lower edge of the roof, where the fascia board lines intersect (Figure 10-18). Note that a precise line is developed along the rafter ends to the old fascia and roof edge. Next, measure a like distance from the last common rafter (or subsequent jack rafters) to a point where the valley

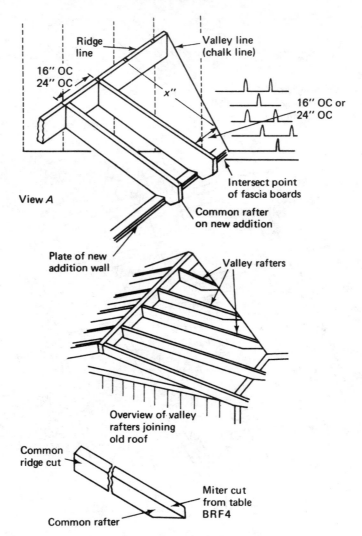

Ridge line

Valley line (chalk line)

16" OC
24" OC

x''

16" OC or
24" OC

Intersect point
of fascia boards

View A

Common rafter
on new addition

Plate of new
addition wall

Valley rafters

Overview of valley
rafters joining
old roof

Common
ridge cut

Miter cut
from table
BRF4

Common rafter

Layout marked on framing square blade (12")
where valley rafter is to be installed over old
roof sheathing

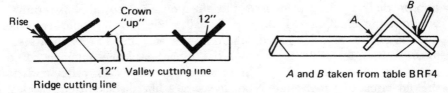

Rise

Crown
"up"

12"

12"   Valley cutting line

Ridge cutting line

A

B

A and B taken from table BRF4

Figure 10-17 Valley Rafter Cut

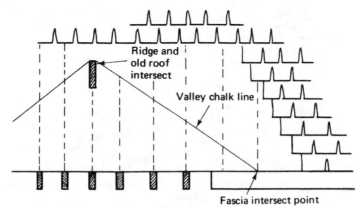

Figure 10-18 Developing the Valley's Line

jack rafter will be nailed to the roof. This point, where it intersects with the valley line measured to the top of the ridge, provides you with a total length of the rafter. Since the valley is equal on both sides of the ridge of the new roof, you can make a left- and right-side valley jack rafter as a pair.

Cut the pair of rafters and install them. Repeat the process described before and measure the next rafter for its length.

Valley Preparation

To prepare the valley, the old shingles must first be removed to expose the sheathing. Then a decision must be made: Can you nail the valley jack rafters directly onto the sheathing, or does some support need to be added? If the sheathing is 3/4 in.-stock sheathing boards, you may elect to nail the rafters directly to the sheathing. If a plywood sheathing 1/2- or 3/8-in. thick is used on the old roof, you should install over the plywood a valley support piece of stock to which you can securely nail the jack rafters. A piece of 1 x 8 or 1 x 10 stock will make an excellent foundation. It should be laid along the valley chalk line previously snapped and securely nailed to each rafter (on the old roof) with a minimum of three 12d common nails per rafter (Figure 10-19).

CHARACTERISTICS OF LEAN-TO ROOFS

As stated in the opening remarks of this chapter, a lean-to roof equals one-half of a gable roof. Its slope, as a rule, is very shallow, and its construction is relatively easy because no ridges are involved and usually no valley rafters are needed.

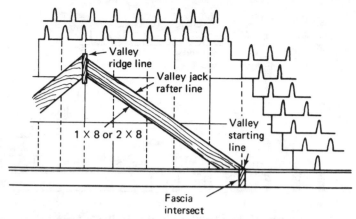

Figure 10-19 Strengthening a Valley Before Rafter Installation

A lean-to roof may stand alone, as for instance when installed on a utility shed, but more likely it will be used to cover a patio or a carport or become the roof for an addition to a home. In the case of the utility shed, the rafters would fit onto the opposing plates and extend over the walls. However, when a lean-to roof is used against an existing home, certain problems arise that must be identified, planned for, and interpreted into tasks.

The lean-to roof will, as a rule, tie into one of two places: an existing roof or the side of the building. Tying into the existing roof is usually required because ceiling clearance below the roof must be maintained and the slope must be sufficient to carry away the rainwater. Tying into the side of a building means exposing wall studs so that a secure fastening is obtained and proper flashing can be installed.

Similiarities to Common Rafters

Each lean-to rafter is very similar to a common rafter. It has a space dedicated as a bird's mouth, frequently marked when not actually cut out. It has a rise per foot of run cut on each end of the rafter: one where the ridge would normally be and one at the overhang. It should be laid out and cut in the same manner as the common rafter, using the shop technique for common rafters.

Tying in to an Existing Roof

Expose the area of the old roof where the new lean-to roof will be installed by removing shingles and sheathing. Measure up from the lower roof line a distance that will represent the lean-to ridge line. Snap a chalk line at this point, parallel to the lower roof line. This chalk line will be the point at which the top to the lean-to rafters will be aligned when they are fastened (nailed) to the old roof rafters.

In conjunction with preparing for tying in the rafter, spacing and positioning will be dictated by the placement of the old roof's rafters. Your only task for aligning them at the outer plate's edge is to verify that the lean-to rafters are at a right angle (90 degrees) to the outer wall's plate. You can do this with your framing square or by using the 3-4-5 method. Once one rafter is aligned, normal spacing methods can be used to mark the positions for the remaining rafters. Figure 10-20 shows a lean-to rafter tied into an existing roof.

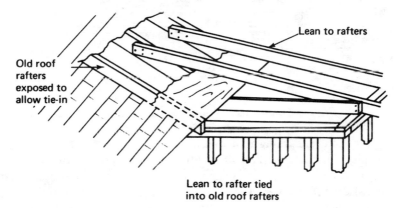

Lean to rafters

Old roof rafters exposed to allow tie-in

Lean to rafter tied into old roof rafters

Figure 10-20 Lean-to Roof to Existing Roof

Tying in to an Existing Wall

This application is only applicable for a two-story house. The lean-to roof will tie into the second-floor studs.

Remove the building's siding and sheathing along the area where the rafters will be installed. Measure up from the new wall's plate area to the height of the top of the lean-to rafter (Figure 10-21). Snap a level chalk line on the exposed 2 x 4s. Precut your rafters by using the shop method for common rafters. Nail one rafter to an exposed stud with the rafter's top edge even with the chalk line. Use a square or the 3-4-5 method to align its opposite end on the plate. Nail, when at right angles, to the plate run. Use standard methods for placing the remaining rafters.

Sheathing and Reinstalling Materials

After all the rafters are installed, the roof can be sheathed. Use Chapter 9 and Routine BS3 as an aid to sheathing. Gable ends should be studded just as the gable roof was studded; use routine BRF5.

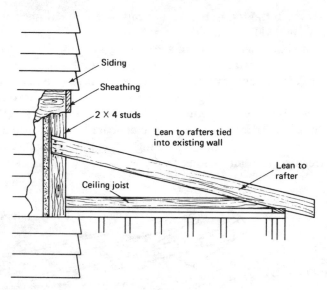

Figure 10-21 Lean-to Roof to Existing Wall

Reinstall the sheathing removed above the top of the lean-to rafters. Install a piece of flashing the full length (may be more than one piece) after first rolling one or two layers of 15-lb felt onto the roof and up the wall. Reinstall the wall's siding. Shingle the roof, whether built onto an existing roof or onto a wall.

ROUTINES

Each of the routines that follows is a complete task within itself.

BRF1: COMMON RAFTER LAYOUT

RESOURCES

Materials:
Total number of rafters needed for the roof: _____ of 2 × _____
                                               no.   width

× _____ rafters
   length (ft)

Building plans with elevation drawings and details
8 to 10 ft 2 x 6

Tools:

1 50-ft tape
1 6-ft folding ruler
1 24-in. level
1 chalk line
1 pair sawhorses
1 framing square
1 plumb bob
1 no. 8 crosscut handsaw

ESTIMATED MANHOURS

Initial layout time   2 hours
Additional layout time   15 minutes per rafter

PROCEDURE

Step 1 Measure and snap a chalk line 12 in. from the shoe at the corners of the span.

Step 2 Mark the center of the span on the chalk line of the wall.

Step 3 Drop a plumb bob from the wall's plate to the center span mark. Mark the plate and snap a chalk line connecting the marks.

Step 4 Optional: Face-nail an 8- to 10-ft 1 x 6 centered on the lines previously made. Snap a vertical center line.

Step 5 From your plan, obtain the rafter rise in inches per foot of run. Multiply this number by the total run in feet and fraction thereof (e.g., 12 ft 6 in. run with 4-in. rise = 12 ft 6 in. × 4 in. = 50 in.).

Step 6 Measure and mark a line from the intersect point of the base line and center-of-span line equal to the product of step 5.

Step 7 Snap a chalk line from the mark made in step 6 to the mark 12 in. up from the shoe at each corner.

Step 8 Partially drive a 16d common nail in the intersect point of the slope line and center-of-span line.

Step 9 Nail a scrap block of 2 x 4 flush with both the outside corner post and the 12-in. mark.

Step 10   Lay a rafter with the crown up on the 16d nail and 2 x 4 block. Tack-nail when the top edge of the rafter passes the center-span line.

Step 11   Mark a plumb line with a pencil and level even with the inner and outer edges of the 2 x 4 block.

Step 12   Draw a level line from the inner 2 x 4 mark and rafter lower edge to the outer 2 x 4 vertical line.

Step 13   Remove the rafter to sawhorses and cut out the wedge-shaped piece with the crosscut saw.

Step 14   Lower the 16d nail at the ridge by the depth of the heel cut. Reinstall the rafter onto the form with the notch area seated on the 2 x 4 block and the upper end on the 16d nail.

Step 15   Use your level to draw a plumb line on the rafter even with the center-span line. Measure in one-half the thickness of the ridge from this line and make a second vertical line. Mark it "cutting line."

Step 16   Overhang layout (optional):
a. Determine the overhang requirements from your plan.
b. Measure on a horizontal plane the overhang distance from the outer edge of the 2 x 4 block.
c. Draw a vertical line.
d. Lay out for a horizontal rafter overhang cut according to plan requirements. Draw a level line from the vertical line toward the bird's mouth.

Step 17   Remove the rafter.

Step 18   Try for fit after the cut. Cut a second rafter.

Step 19   Try for fit on the opposite side of the span. If it fits, use the rafter as a pattern to cut all rafters required.

BRF2: GABLE-FRAME ROOF ASSEMBLY

RESOURCES

Materials:
Ridge member or members
Roof layout plan
Precut rafters

1/2 lb 12d or 16d common nails per rafter set
2 x 4s for braces

Tools:
2 12- to 16-oz claw hammers
1 plumb bob
1 framing square
1 24-in. level
1 6-ft ruler or tape
1 6- to 8-ft straightedge
1 pair sawhorses
1 12-ft or extension ladder
1 no. 8 crosscut handsaw

ESTIMATED MANHOURS

For two people   1.5 hours per 8 lineal ft of roof

PROCEDURE

Step 1 Lay out the rafter position on opposing wall plates (16 or 24 in. OC). For maximum strength, try to have the rafters rest above the common studs.

Step 2 Mark the ridge board similar to the plate. Cut the ridge on the center of a rafter if more than one piece of ridge is needed.

Step 3 Lay all precut rafters against the wall with their ridge cuts up.

Step 4 Install the walk surface on the joists where the ridge will be located.

Step 5 Nail two rafters on the same side of the ridge member at opposite ends.

Step 6 Have one man raise the ridge until the heel cut on the rafter fits tightly onto the plate. Nail to the plate.

Step 7 Repeat step 6 for the other two rafters.

Step 8 Install two more rafters on each side of the ridge.

Step 9 Install a 2 x 4 brace from a point on the ridge between the first and second rafters at approximately a 45-degree angle back from the outer gable end wall and tack-fasten to the joist:
a. Drop a plumb bob from the end rafter at the ridge and align the bob with the outer edge of the gable-end wall. Rack the roof either in or

out to align the bob. Tack-nail the 2 x 4 to the joist when the bob is in line.

b. Alternative: a straightedge and level may be substituted for the plumb bob and line.

Step 10   Install the remaining rafters in sets to keep the ridge line straight.

Step 11   Individually cut and install jack studs in the gable end, aligning them over the common wall studs below. Use a level to plumb the stud while marking each for length and cutout. See BRF5 for a sample jack-stud layout.

Step 12   Permanently install a brace at each end of the ridge and collar beams as in Routine BRF3.

Step 13   If you have not previously marked and cut rafter ends, use the BRF1 instructions.

Step 14   Sheath the roof according to Routine BS3 in Chapter 9.

BRF3: ROOF BRACING AND COLLAR BEAMS

RESOURCES

Materials:

_____ of 2 x 4 stock for each brace
  lineal ft

_____ of 1 x 6 stock for each collar beam
  lineal ft

1 lb 12d common nails for 2 x 4 brace installation
1 lb 8d common nails per collar beam

Tools:
1 6-ft folding ruler
1 24-in. level
1 combination square
1 framing square
1 13- to 16-oz claw hammer
1 chalk line
1 no. 8 crosscut handsaw

1 plumb bob
1 straightedge

ESTIMATED MANHOURS

For a 2 x 4 brace   30 minutes
For a 1 x 6 collar beam   20 minutes

PROCEDURE A: 2 x 4 BRACE

Step 1 Precut one end of the 2 x 4 brace as shown in Figure 10-14A. The first mark is made at a 45-degree angle using your combination square. The second mark is made at 90 degrees from the first.

Step 2 Lay the base of the cut end of the 2 x 4 on a 2 x 4 nailed across two or more ceiling joists, or on the gable end plate.

Step 3 Mark the free end of the 2 x 4 flush with the top edge of the ridge board. Cut along the mark.

Step 4 Verify that the gable end is still plumb by using your level and straightedge or plumb bob. Adjust if necessary.

Step 5 Install the brace with nails.

PROCEDURE B: COLLAR BEAM

Step 1 Measure the length of the collar beams needed by using your roof plan or actual measurement from rafter to rafter.
a. On plan: measure down 18 in. (as per scale) from the ridge and along the run (as per scale). Multiply by 2. The product is the total length of the collar beam.
b. On roof: use a ruler, straightedge, and level to measure down 18 in. Mark the position of the collar beam on a rafter set.

Step 2 Lay out the collar beam (1 x 6) and bevel each end of the beam by using the framing square and the roof's rise per run.

Step 3 Cut at one time all pieces needed.

Step 4 Nail the first collar beam in place using at least four 8d common nails per beam end.

Step 5   Measure down from the ridge a point equal in length to the bottom edge of the first beam installed.

Step 6   Place a mark at the other end of the roof equal in length to the mark measured in step 5. Snap a chalk line.

Step 7   Repeat steps 5 and 6 on the opposite side of the ridge.

Step 8   Install the remaining collar beams.

BRF4: VALLEY JACK RAFTER LAYOUT

RESOURCES

Materials:
_____ of 2 x _____ valley rafter stock
  lineal ft   width

_____ lb 12d or 16d common nails

_____ lb 8d common nails

2 1 x 8 x _____ lineal ft.

Stock to line the valley to provide a base on which to nail the rafters (optional).

See Procedure A on page 161.

Tools:
1 6-ft folding ruler
1 framing square
1 bevel square
1 combination square
1 no. 8 crosscut handsaw
1 14- to 16-oz claw hammer
1 25-ft tape measure

ESTIMATED MANHOURS

20 to 30 minutes per rafter

PROCEDURE A: ESTIMATING THE MATERIALS NEEDED FOR VALLEY JACK RAFTERS

Step 1 Using your plan (and its scale), measure each rafter needed. Add convenient lengths to determine the most economical size of stock to purchase. Put these data into the programmed plan for this task.

Step 2 *Alternative:* lay out the rafter placement on the ridge. Use a tape to measure each rafter for length.

PROCEDURE B: VALLEY JACK RAFTER

Step 1 Lay out the rafter spacing on the ridge (16 or 24 in. OC).

Step 2 Draw lines on both sides of the ridge where previous spacing marks are located.

Step 3 Determine the intersect point of the new roof and old roof by laying a straightedge along the new common rafter ends. Mark and partially drive a nail into the point.

Step 4 Snap a chalk line from the nail driven in step 3 to the top end of the ridge where it ends on the old roof.

Step 5 (Optional) Install a 1 x 8 along the chalk line from the intersect point to the ridge (on the ridge side of the chalk line).

Step 6 Measure an interval (16 or 24 in. OC) from the last common rafter (parallel to the ridge) and mark the old roof's sheathing or 1 x 8.

Step 7 Measure and record the total length of the first (and subsequent) valley jack rafter.

Step 8 Cut the common rafter ridge cut on the member that will be the valley jack rafter.

Step 9 Measure from the long point of the ridge cut the total length of the valley jack rafter (see Figure 10-17).

Step 10 Lay out the heel cut from the point just made by laying the framing square on the rafter with the base on the tongue and 12 in. on the blade on the lower edge of the rafter, as shown in Figure 10-17. Have the outer edge of the blade intersect the rafter length point made in step 9.

Step 11 Lay out the rafter side cut by using Table BRF4 to obtain the proper bevel.

Step 12 Simultaneously cut both marks made on the rafter. Try for fit.

Step 13 (Optional) The cut rafter may be the pattern for the opposite rafter set.

Step 14 Repeat steps 6 through 13 for each succeeding valley jack rafter.

Optional: The third and fourth lines of the rafter data tables on the framing square may be used in lieu of step 7 for calculating the length of subsequent valley jack rafters. The values listed on the square equal the difference in length from one rafter to the next.

TABLE BRF4: Valley Jack Rafter Side Cuts

Rise per Foot (in.)	Setting on Framing Square	
	Tongue	Blade
3	7-3/4	8
4	7	7-1/4
5	16	17
6	15	16
7	12	13
8	9	10
9	7	8
10	13	15
12	9	11
14	7	9
16	8	11
18	11	16

Set the framing square for any combination of numbers needed according to the rise per foot of run of the roof. Mark along the blade of the square (24-in.-length or 2-in.-width side).

BRF5: LAYING OUT AND INSTALLING GABLE-END STUDS

RESOURCES

Materials:

_____ lineal feet 2 x 4 for jack studs. The total number of studs
　no.

needed for one rafter run × the longest stud needed + 10 percent equal lineal feet required.

Tools:
1 bevel square
1 6-ft ruler
1 13- to 16-oz claw hammer
1 no. 8 crosscut handsaw
1 combination square
1 24-in. level
1 1-1/2-in. chisel
1 5-1/2-point ripsaw

ESTIMATED MANHOURS

4 to 5 hours per average gable end

PROCEDURE

Step 1 Mark the top side of the plate (over the common stud) where each gable end stud will be installed.

Step 2 Plumb the level from the first stud to the rafter. Make a pencil mark on the rafter.

Step 3 Measure and precut a piece of 2 x 4 long enough to *almost* reach the top side of the rafter.

Step 4 Place the stud on the plate and behind the rafter, also on the pencil lines. Mark the stud for a bevel cut.

Step 5 Set the combination square for 1-1/2 in. and mark the stud from the bevel mark to the end of the stud (the part behind the rafter).

Step 6 Cross-cut at the bevel and make at least two additional cuts to a depth of the 1-1/2-in. line.

Step 7 Rip or chisel the stock away as shown in Figure 10-15.

Step 8 Nail in place.

Step 9 Lay out and cut the second stud by repeating steps 2 through 8.

Step 10 Measure and record the difference in length between the first and second studs.

Step 11 Lay out the rafter cut on all remaining studs required and add one additional length difference for each succeeding stud up to the ridge.

Step 12 Repeat this process for the other rafter area *but* decrease the length of each succeeding stud.

BRF6: LAYING OUT AND INSTALLING A LEAN-TO ROOF

RESOURCES

Materials:
_____ 2 × _____ × _____ rafters
    no.   width   length

_____ lb 12d or 16d common nails. 1/2 lb nails are needed per
    no.

rafter.
Rafter layout from plans
1 roll tarpaper and nails

Tools:
1 no. 8 crosscut handsaw
1 chalk line
1 25-ft tape
1 6-ft ruler
1 ripping bar
1 crowbar
1 12-ft ladder

ESTIMATED MANHOURS

For two people   4 to 8 hours to expose rafters or studs
                        20 minutes per rafter layout and cut
                        30 minutes for rafter placement layout
                        20 minutes per rafter installation

PROCEDURE

Step 1 Lay out the rafter as per plan for the heel cut (if used).

Step 2 Remove the shingles and sheathing or siding and sheathing where the new rafters will be tied into the old structure.

Step 3 Tie one end of a chalk line (or mason line) to a scrap piece of rafter stock tacked on top of the plate.

Step 4 Stretch the line from the nail onto the roof or wall at a height needed for the proper slope measured from the plate as shown in Figures 10-20 or 10-21.

Step 5 Mark the point on the roof or stud wall where the line touches.

Step 6 Snap a horizontal line from the point made in step 5 across the area where the rafters will be installed.

Step 7 Nail rafters, crown up, to each old rafter or old stud with 16d common nails. Use four to five nails per rafter.

Step 8 Position and nail the rafters to the plate.

Step 9 Trim the overhang of the rafters according to Routine BRF1 procedures.

Step 10 Cut and install jack studs between the plate and the end rafter, using the instructions in Routine BRF5.

BRF7: INSTALLING GABLE-END OVERHANG RAFTERS

RESOURCES

Materials:
2 2 x 4 × _____ fly rafters per gable end
        ft

_____ of 2 x 4 × _____ for lookouts. The length of a lookout
  no.   lineal ft

equals the desired overhang in inches less 1-1/2 in. plus the distance from the outside surface of the end rafter to the same surface on the second rafter.

Detail portion of the plan illustrating the roof cornice design

5 lb 16d common nails per 32 lineal ft of fly rafter installation

Tools:

1 no. 8 crosscut handsaw
1 16-oz claw hammer
1 3/4- or 1-in. wood chisel
1 framing square
1 extension ladder
1 adjustable square
1 6-ft folding ruler
1 scaffolding (built on the site from 2 x 4s and 1 x 6s) (optional)

ESTIMATED MANHOURS

For two people   4 hours for 16 lineal ft of fly rafter

PROCEDURE

Step 1 Lay out the end rafter on the gable end on 24-in. centers.

Step 2 Continue layouts by developing a pattern for a 2 x 4's thickness and to a depth of 3 to 3-1/2 in.

Tip: On a 2 x 6 rafter, notch only 3 in. and notch the lookout 1/2 in. On a rafter wider than 2 x 6, cut the full 3-1/2 in. for the lookout.

Step 3 Measure and precut *all* the lookout pieces of 2 x 4s needed according to the formula: overhang less 1-1/2 in. + distance from gable-end rafter to second rafter.

Step 4 Mark a line on all lookouts equal to the overhang length less 1-1/2 in.

Step 5 On the second rafter, make a set of 24-in.-OC marks at the same place as the gable-end rafter.

Step 6 Install metal straps if used and/or nail lookouts into position with 16d common nails.

Step 7 Precut the ridge cut on the fly rafter.

Step 8 Nail the fly rafter in position with the ridge cut in place and the tops of the lookouts and fly rafter even.

Step 9 Trim the lower fly rafter end even with the other rafter ends.

Step 10 Repeat steps 6 through 9 for the opposite roof slope.

11

SHINGLING

BASIC TERMS

Interleaving method of applying strip shingles in a valley where the shingles from one roof pass through the valley onto the adjacent roof.

Square 100 sq ft of roof area; three bundles of shingles are sufficient to cover 100 sq ft of roof.

Shingling a roof with composition strip shingles is a relatively simple task. Two factors distinguish a good from a bad job: (1) the vertical and horizontal alignment of the shingles, and (2) the proper nailing method.

When working on any roof surface, there is a certain amount of danger. On roofs whose pitch is 6 in. per foot of run or less, there is little requirement for scaffolding, but on roofs with pitches of 8 or 10 in. or more rise per foot of run, a scaffold is needed. Scaffolding is discussed in detail later.

In addition to shingle application, there may be occasion to flash a joint area with metal to prevent leaks. A routine in addition to the three shingling routines is provided at the end of the chapter to outline flashing installation.

SCOPE OF THE WORK

The work involved in shingling a roof with strip shingles is three-fold. The flat (sloped) areas are shingled in a precise manner involving horizontal and vertical alignment. The chalk line and the shingle itself are used to maintain these lines. The work is fast and easy.

The second area is the valley. It requires a special application of shingles. The method described later is one of *interleaving* shingles from both roof surfaces across the valley. This method replaces the older method of installing a metal valley. The third and last area is the ridge cap. Flashing made from aluminum or copper strips 12, 16, 18, or 24 in. wide is used to waterproof such areas as around chimneys or when a lean-to or gable roof intersects the side wall of a house.

Before explaining installation characteristics, the shingle and its properties must be understood. With these facts you will more readily complete the calculations and understand the manhour allocations listed in the routines that follow.

SHINGLE MAKEUP

Composition shingles are made from an asphalt- or vinyl-impregnated paper (tarpaper) with aggregate stone pressed into the exposed surface area. Note in Figure 11-1 that the aggregate (crushed stone) is concentrated on the lower portion of the shingle, which is the surface area exposed to the weather. Also note that this area measures approximately 5-1/2 in. from the shingle's lower edge. Customarily only 5 to 5-1/4 in. are exposed to the weather. This means that 7 in. of the shingle is overlapped by the next row of shingles.

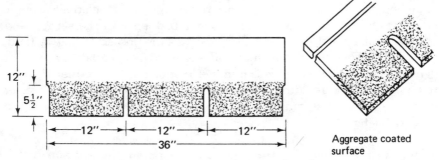

Aggregate coated surface

Figure 11-1 Typical Strip Shingle

Industry has provided a securing technique which, to a fair degree, prevents shingles from being torn loose by winds. Spots of tar/asphalt are strategically placed on the underside along the outer-edge surface of the shingles as shown in Figure 11-2. This sealing, plus the overlap, provide a secure foundation for each shingle on the roof's shingle surface.

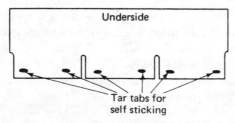

Tar tabs for self sticking

Figure 11-2 Self-Sticking Shingles

You should buy materials that are rated at 220 to 240 lb. per square (there are better selections at 240 to 290 lb per square). A square of shingles equals the number of shingles needed to cover 100 sq ft of roof area (three bundles equal one square).

TYPICAL CUTS IN SHINGLES

Refer again briefly to Figure 11-1. Note the dimensions of the shingle. Its total length is 36 in. and width 12 in. Each cutout is centered at 12 in., one-third the length of the shingle. Also note that one-half a cutout is made on each end of the shingle. Study Figure 11-3 so that when you become familiar with the way shingles are installed, you will understand some of the typical cuts made in strip shingles.

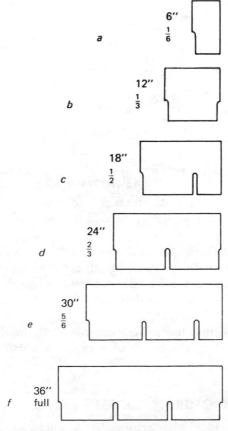

Figure 11-3 Typical Cuts for Strip Shingles

ESTIMATING SHINGLE REQUIREMENTS

Determine the square footage of roof to be covered. If you used Routine BS3, Sheathing a Roof, you can obtain from it the total number of square feet of sheathing used to cover the roof.

If you are reroofing an existing roof, you must measure each slope for its length and width. Multiply the length × width and obtain the product. Then double your figure to account for both slopes (Figure 11-4).

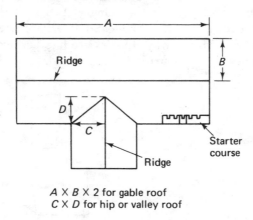

$A \times B \times 2$ for gable roof
$C \times D$ for hip or valley roof

Figure 11-4 Calculating Shingle Requirements

Valley and hip areas require the same number of shingles. Since each half (slope) is equivalent to a right triangle, simple multiplication of its length × width will result in the total requirement for both slopes. Since there will be some cutting and waste, add 1 bundle of shingles for each 16-ft-long valley.

Next, measure the total lineal feet of ridge. Include the hip ridge lengths. Divide the total number of lineal feet determined by a value of 27 to find the total number of bundles needed to cover the ridges.

Finally, determine how many bundles of shingles are to be used as starter-row shingles. These are shingles installed upside down as the undercourse for the first layer. Measure the total lineal footage of the lower roof edge and divide by a value of 78 to obtain the number of bundles needed.

SHINGLE EXPOSURE AND ALIGNMENT

Recall that a shingle has crushed stone impressed into the surface and that the exposed area is limited to 5 or 5-1/4 in. This will ensure that the joint between shingles is covered adequately and that the shingle has maximum hold down strength.

Three considerations must be employed simultaneously: (1) perfect horizontal alignment of the bottom edge of the row of shingles, (2) perfect vertical alignment of shingle cutout areas, and (3) a 5-in. exposure.

A chalk line and ruler are the best tools to use to keep the shingles horizontal, vertical, and spaced properly. The two roof edges needed as guides are the lower roof's edge and the gable end (if one exists). On a hip roof, snap a vertical line by using the 3-4-5 method of squaring a right angle (see Chapter 10 for an explanation of this method).

The horizontal lines should be snapped for every fourth or fifth row of shingles. The first one, shown in Figure 11-5, is calculated as follows. The first row of shingles will extend 1 in. over the lower edge of the roof. Snap a chalk line 11 in. up on the roof's slope. Next, add to this the exposure for three more rows of shingles (15 in.) and snap a chalk line at 26 in. from the roof's lower edge. Snap a series of lines at 20-in. intervals above the second chalk line.

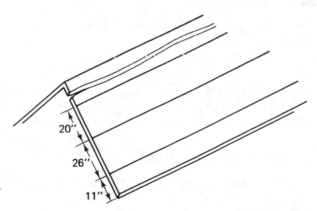

Figure 11-5 Snapping Horizontal Chalk Lines

Using the sloping roof edge, measure in 5-1/4, 11-1/4, 17-1/4, 23-1/4, 29-1/4, 35-1/4 in. at the ridge and lower roof edge (Figure 11-6). Snap a series of lines connecting the measurement points, thereby creating vertical lines. The 3/4-in. overhang that results is necessary to shed the rain away from the sheathing.

Apply the same principle of vertical lines to any hip area after first snapping a vertical line at 90 degrees from the lower roof edge.

Shingle alignment of ridge capping and hip capping is easily accomplished by positioning a one-third shingle (Figure 11-3) over the ridge with half the shingle's width on each slope.

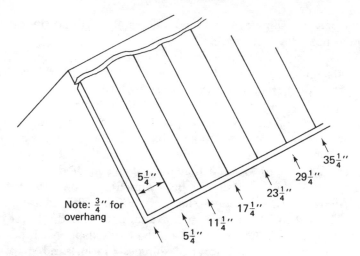

Figure 11-6 Vertical Lines for Shingle Alignment

NAILING SHINGLES

Strip shingles should be nailed with 1-in. hot-dipped galvanized roofing nails. The preferred method requires six nails per shingle (Figure 11-7). Note in Figure 11-7A that a nail is placed 1 in. to each side and above each slot. Also note that the height of the nail from the lower edge of the shingle is 6-1/2 in. It is important that you nail at the proper height to ensure that each nail will be covered by the next course and will be driven through the shingle and the one under it, then into the roof.

APPLICATION OF SHINGLES

A layer of 15-lb felt tarpaper is usually installed over the sheathing and nailed securely.

A starter row (refer to Figure 11-8) made with standard strip shingles (nailed upside down) must be installed before the regular shingles can be installed. This starter row is a must because it provides the basis for a double row of shingles.

Precisely cut length of shingles (Figure 11-3) to begin each new row. This ensures the breaking of joints at different places from the row below it, thereby preventing leaks. Use the vertical and horizontal lines snapped on top of the tarpaper as your guides. Between the horizontal lines use the shingle's cutout or a 5-in. scrap piece of wood as a guide for positioning succeeding rows.

The shingle at the ridge (not the ridge cap) must lap over the ridge *and* its notches must measure 6 in. or less from the ridge (Figure 11-9).

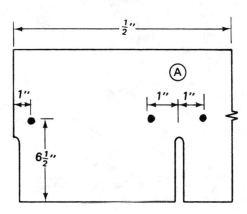

Figure 11-7 Where to Nail Shingles

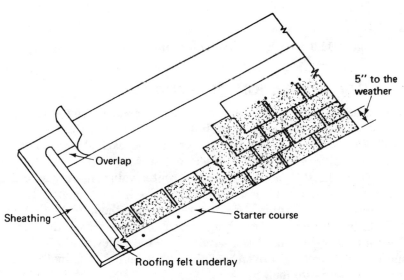

Figure 11-8 Shingling a Slope

You may use a precut ridge piece of shingle to see if the last row of shingles is high enough to prevent leaks.

Repeat the process detailed above for the opposite slope, then apply the ridge cap shingles as shown in Figure 11-9. Maintain a 5- to 5-1/4-in. exposure, and fasten one nail on each slope about 6-1/2 in. up from the shingle's lower edge. Top-nail the final ridge cap piece with four galvanized nails.

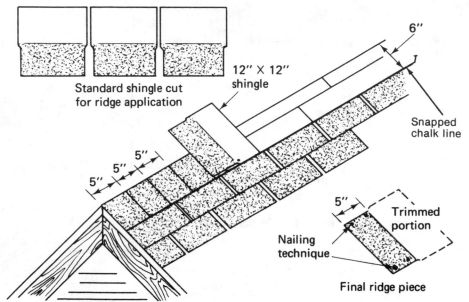

Figure 11-9 Last Row of Shingles and Ridge-Cap Shingles

VALLEY APPLICATION OF SHINGLES

In order to accomplish the task, both roof areas joining at the valley must first be shingled to the valley (but not into the valley). At this point the valley can be laid. In order to be watertight, each shingle must pass through the valley and onto the other roof surfaces. To ensure this, the top corner of the shingle must pass the center valley line by 1 to 2 in. The excess shingle should be trimmed.

Figure 11-10 illustrates the interleaving method. Note that each shingle butts against the last shingle installed and passes through the valley. Note that the shingle is *creased* along the valley line. With it held firmly in place, a nail is driven into the upper corner and another lower down on the opposite roof.

The second shingle may be installed from the same roof through the valley, or from the opposite roof through the valley. Try a shingle from the other roof first. If it provides excellent coverage where, for instance, you may have had to join shingles in the valley, install the shingle. On the other hand, if a joint was made, it might be better to install a second shingle (5 in. higher) on the same roof to cover the joint before bringing a shingle through the valley from the other roof. Repeat the process until the valley(s) is complete.

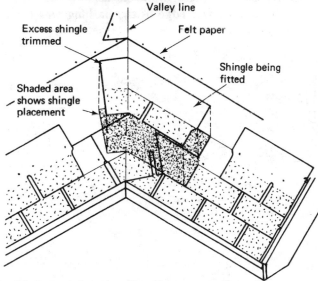

Figure 11-10 Interleaving Shingles in a Valley

JOINING A LEAN-TO ROOF'S SHINGLES TO A MAIN ROOF'S SHINGLES

Joining a lean-to roof's shingles requires that the lean-to roof's shingles pass under the main roof's shingles. Therefore, the nails must be removed from the two courses of shingles on the main roof just above the joining point of the two roofs. This will allow full-width shingles from the lean-to roof to be installed. Once positioned and nailed, the main roof's shingles, previously loosened, may be renailed in place.

FLASHING

Flashing, metal installed in various places to prevent leaking, is required wherever a roof joins another object such as a chimney or siding on a house wall. Since details in this book provide for building a lean-to roof alongside an existing wall, Routine BRS4 is provided (at the end of the chapter) as an aid to flashing the roof. In this application a strip of flashing metal 14 in. wide and the length of the roof is purchased from a lumberyard.

Once unrolled, it is formed into a angle approximating the angle where the roof joins the house wall. Figure 11-11 shows where the flashing is installed and nailed, and how it looks when bent. Note that the nails joining the side wall are placed high up on the flashing. Also note that the flashing is pressed into an asphalt sealing mixture that was applied before the lower edge of the flashing was *top*-nailed into the roof.

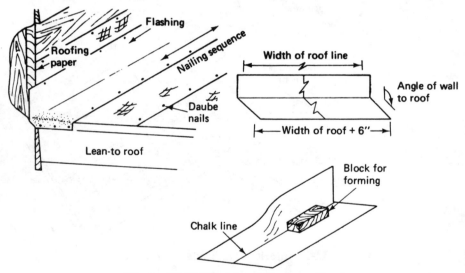

Figure 11-11 Flashing a Lean-to Roof

The tops of the nails may also be dabbed with tar to waterproof any holes made by them. Finally, note that the flashing is allowed to extend past the roof's edge to prevent blowing winds from driving rain behind the flashing.

USING ROOF JACKS ON A SHINGLING JOB

The pitch of a roof is often too steep to shingle without some safety precautions being taken, one of which is use of a special roof jack (Figure

11-12). Two 8d common nails are driven through the slots at the upper end and into the *rafters. You must make sure that the nails are into the rafter.* Install two jacks spaced 8 to 10 ft and lay a 2 x 6 member in the space provided. The 2 x 6 should extend 1 ft past each jack.

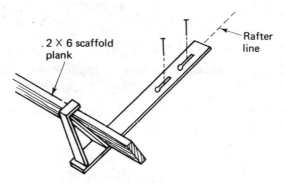

Figure 11-12 Roof Jacks

Figure 11-13 shows a ladder scaffold jack. If you use this type of jack on your extension ladders, you need to rent a pair. When installed, the top must be either horizontal or sightly sloped toward the house. If ladder jacks are used, two 2 x 10 planks should be used as your scaffold and a scrap piece of 1 x 4 or 1 x 6 should be nailed across the center of the 2 x 10s. This will force both 2 x 10s to bend together when walked upon. If the piece is not installed, each 2 x 10 will bend separately and unevenly, creating a severe hazard.

Caution: When using scaffolding, observe extreme caution when installing the scaffolds, installing the scaffold planks, and working on the scaffold.

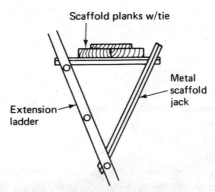

Figure 11-13 Ladder Jacks

ROUTINES

The following four routines provide the aids needed to do almost every type of roof. Each one is detailed to provide data and instruction on a phase of the total job. These tasks may be employed to shingle a new roof surface or to reshingle an old roof surface. Within the procedures of Routine BRS1, Strip-Shingle Application, instructions for laying tarpaper are included. Routine BRS2, Shingling a Valley, and BRS3, Shingling a Ridge, provide details about these special applications. Routine BRS4, Flashing a Lean-to Roof, provides guidelines for roof flashing.

BRS1: STRIP-SHINGLE APPLICATION

RESOURCES

Materials:

＿＿＿＿＿＿ of shingles. One square of shingles equals 100 sq ft of roof.
no. squares

(*Note:* Use your estimate of roof sheathing to calculate the number of squares needed.)

＿＿＿＿＿＿ of rolls of 15-lb felt tarpaper. One roll covers about 400 sq ft of roof.

＿＿＿＿＿＿ lb 1-in. galvanized roof nails. (*Note:* Nails should penetrate sheathing by 1/4 to 1/2 in.)

Tools:
1 chalk line
1 13-oz claw hammer
1 roofing, linoleum, or wallboard knife
1 6-ft folding ruler
1 framing square
1 extension ladder
1 or more pairs of roof jacks for high-pitched roofs (optional)
1 or more pairs of ladder jacks with which to start roof (optional)
1 or more 2 x 6 x 12 ft to be used as scaffold boards (optional)

ESTIMATED MANHOURS

Tarpaper-roll installation	1 hour per roll
Chalk-line la, out average	30 minutes
Installation of shingles	2 hours per square

PROCEDURE

Preliminary step 1 Roll out the tarpaper and cut off in 10- and 12-ft lengths.

Preliminary step 2 Lay the first strip of paper flush with the end of the roof and the lower edge. Nail the upper edge each 12 to 15 in. Nail the lower edge each 2 ft.

Preliminary step 3 Install the second and remaining strips of tarpaper by overlapping the first sheet 3 in.

Preliminary step 4 Install the second row by breaking joint with the first row, and overlap the first row by 2 in. Nail.

Step 1 Snap the following lines vertically on the roof and parallel with the roof's edge: 11-1/4 in., 17-1/4 in., 23-1/4 in., 29-1/4 in., and 35-1/4 in.

Step 2 Snap chalk lines parallel with the lower roof's edge at 11, 26, and 20 in. thereafter.

Step 3 Position your starter-row shingle upside down to overhang the roof's end by 3/4 in. and lower the roof's edge by 3/4 in. Nail it 6-1/2 in. up from the bottom edge with six nails.

Step 4 Install a row of shingles over the starter row. Be sure to keep your nails above the notches by 3/4 to 1 in.

Step 5 In the second row, trim 6 in. from the end of a shingle that will project over the roof's vertical edge.

Step 6 Align the shingle's end with the 30-in. chalk line and nail it in place 5 in. above the lower edge of the first row. Use your ruler to measure the 5 in.; or cut a 5-in. block.

Step 7 Butt the next *full* shingle against the first, and nail it after positioning it for the 5-in. position.

Step 8 To begin the third row, cut another shingle 24 in. and install it.

Step 9 To begin the fourth row, cut another shingle 18 in. long and nail it with its top edge even with the first chalkline and end with the 18-in. vertical line.

Step 10 Continue laying shingles, keeping in mind that the top of the shingle of each fourth row must align with the horizontal chalk line.

BRS2: SHINGLING A VALLEY

RESOURCES

Materials:
None required, because they were included in the estimate for Routine BRS1

Tools:
1 utility knife
1 13-oz claw hammer
1 6-ft folding ruler
1 pry bar

ESTIMATED MANHOURS

4 hours per 8 ft of valley

PROCEDURE

Step 1   Butt the first shingle against the last full shingle installed on the common roof area.

Step 2   Force it into the valley, creasing the shingles. Make sure that the top corner of the shingle extends past the valley. If it extends past the valley excessively, trim it off. Nail in place.

Step 3   Install the next shingle from the opposite common roof into the valley according to steps 1 and 2.

Step 4   Install shingles alternately up the valley following steps 1, 2, and 3.

Note: In some cases, two rows of shingles may be installed from one common roof area before going to the other roof area.

BRS3: SHINGLING A RIDGE

RESOURCES

Materials:
_____ of bundles of shingles. One shingle covers 15 lineal in. of ridge,
    no.

and one bundle covers approximately 26 lineal ft of ridge (may already
be calculated in Routine BRS1).

    1 lb of 1-1/2-in. galvanized roof nails per bundle of shingles

Tools:
1 chalk line
1 13-oz claw hammer
1 utility knife
1 framing square
1 6-ft folding ruler

ESTIMATED MANHOURS

1.5 hours per 25 ft of ridge

PROCEDURE

Step 1 Cut a bundle of shingles in three pieces 12 in. x 12 in. per
shingle if more than 25 ft of ridge will be covered.

Step 2 Measure down 6 in. from the ridge at two points and snap a
chalk line.

Step 3 Nail the first ridge piece even with the chalk line and roof edge.

Step 4 Lay the next piece over the first and 5 to 5-1/4 in. back, and
nail it flush with the chalk line.

Step 5 Install 5 to 6 ft of ridge pieces along the chalk line, exposing
5 to 5-1/4 in. per shingle.

Step 6   Fold down each shingle over the ridge and nail on the opposite roof area.

Step 7   Install the last piece of the ridge by top-nailing it in place.

BRS4: FLASHING A LEAN-TO ROOF

RESOURCES

Materials:
_____ ft of 14-in. aluminum or galvanized flashing
  no.

1 lb 3/4-in. galvanized roofing nails

Tools:
1 pair snips
1 16-oz claw hammer
1 folding ruler
1 piece 2 x 4 3 to 4 ft long

ESTIMATED MANHOURS

2-1/2 hours per 16 lineal ft

PROCEDURE

Step 1   Measure and cut a piece of flashing 6 in. longer than the width of the roof where it ties into the house wall.

Step 2   Crease the flashing through its entire length as follows:
a. Snap a chalk line where the bend will be made.
b. Lay the 2 x 4 along the chalk line and manually draw the flashing up to crease it.
c. Firm the crease by tapping lightly along the bend with a hammer.

Step 3   Apply a good coat of paste asphalt along the wall and shingle area where the flashing will cover.

Step 4 Position the flashing against the wall and onto the shingle surface, overlapping the lean-to roof edges by 3 in.

Step 5 Start nailing the flashing to the wall from the *center* of the span both left and right at 8- to 10-in. intervals.

Step 6 Nail the lower edge of the flashing to the roof and dab the nails with asphalt.

Step 7 Fold the overhanging metal down slightly to stiffen the edges and prevent blowing rain from entering the corners.

Post step Install wall siding that was previously removed to allow access to the wall's studs.

12

CORNICE

BASIC TERMS

Fascia part of a cornice that is nailed across the rafter ends and onto a fly rafter.

Frieze board fit and nailed between the last course of siding and either lookouts or roof's edge.

Lookout 2 x 4 structural member nailed horizontally from the lower rafter edge overlay to the wall's outer surface.

Soffit underside stock used in closed or boxed cornice extending from fascia to frieze.

When you build the cornice along the eaves and rakes of the house, you are installing the final trim. The work is exacting, and the materials selected for the job must be of first quality. If this is your first experience in building a cornice, you may waste some material. But if you will study the various aids provided in this chapter, you may avoid the added cost.

SCOPE OF THE WORK

Three types of cornice are used on homes today: *flush, open rafter ends,* and *boxed.* There are also numerous variations of these basic types. Each type requires a considerable dedication in time and material. The flush type requires the least amount of time, the open rafter next, and the boxed cornice the greatest amount of time and also of material.

The work will involve considerable climbing from ground to ladder or scaffold. Two people will be needed for much of the work because 12- to 16-ft lengths of lumber are difficult to handle alone.

Frequently, a cornice return is or must be incorporated in the design. The work is exacting and time-consuming. A cornice is usually made from more than one piece of stock.

On roofs with overhangs, lookouts must be cut and installed. This job is lengthy because one lookout is installed on each rafter. Toe nailing should be used as well as face nailing.

Ventilation must be incorporated into the cornice. If commercial soffit material is purchased, the cutouts are already made. If you cut the soffit and/or frieze board (in some installations) for ventilation, the task will take some time.

Since this is finish work, each joint must be cut true to its dimension. All miters must fit very well. Finishing nails must be used, and their heads must be set so that they may be filled during painting. Sanding will be done at the corners and on returns where scallops and coves are cut into stock.

Considerable climbing will be required because each piece must be individually cut to fit. Saw horses with scaffold planks may make the work easier. Stepladders or extension ladders may be needed.

FLUSH (CLOSE) CORNICE

The *flush* or *close cornice* is designed where the rafter ends are cut flush with the wall's framing (blocks in block construction, Figure 12-1). The sheathing on the outside of the wall is extended to the top edge of the rafter. The siding is installed to approximately wall plate height, and a frieze board is formed and fitted to lap over the last course of siding by 1/2 in., extend to the rafter's top edge, or extend to the top of the roof's sheathing; and to return on the corners of the gable end's roof with the gable-end frieze boards.

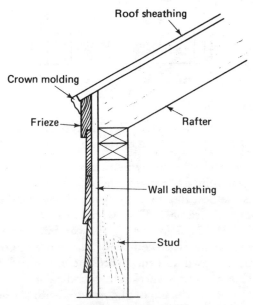

Figure 12-1 Flush Cornice

A frieze board for a horizontal installation differs from a gable end, where the gable-end piece is usually a standard 5/4 x 4 in., 5/4 x 6 in., or 5/4 x 8 in. The horizontal frieze does two jobs at once: (1) it must provide a sealer from the last course of siding, and (2) it must completely cover the sheathing that provides a finishing surface between the wall and the rafter's or roof sheathing's top edges.

One of two methods can be used to properly seal a wall from weather with the frieze board. The board should extend below the last course of siding a minimum of 1/2 in. (Figure 12-2). Figure 12-2A shows a rabbet cut in the lower back side of the frieze with the frieze fitted flush to the sheathing. Figure 12-2B shows blocking installed flush to the siding (most applicable to brick veneer) and the frieze board with no rabbet overlapping the siding but nailed to the blocking.

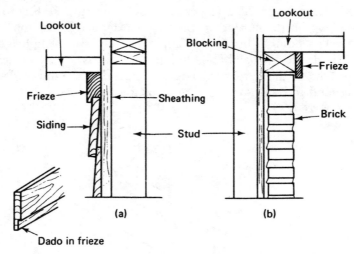

Figure 12-2 Frieze-Board Preparation

Splicing is usually necessary when installing cornice. The 45-degree miter cut is the best splicing joint to use except at gable ends. The combination square is the best tool to use when laying out the stock to be cut.

Assume that an internal splice is required on a 1 x 8 frieze (or fascia) board. Figure 12-3 shows that the combination square with its built-in 45-degree feature is placed along the top edge of the board so that when the line is drawn and the board is cut, the end grain of the cut faces outward. This method assures a perpendicular face cut.

The joining piece of frieze must be mitered with a 45-degree (closed) cut. This cut, also shown in Figure 12-3, again assures a perpendicular face cut and a perfect fit.

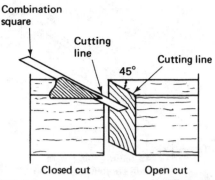

Figure 12-3 Laying Out Splices on Cornice Material

The need to splice trim material on homes is essential to a quality job because stock lumber and molding are subject to swelling and shrinking. If the lumber is purchased with standard 10 to 19 percent moisture content, the adjustments will be 1/16 to 1/8 in. If material is allowed to accumulate water before it is installed, more swelling and shrinking will result. With the use of a miter joint in a splice, material adjustments may be made and the results will not be seen. However, as shown in Figure 12-4, if butt joints are used, gaps will continue to occur in the joints.

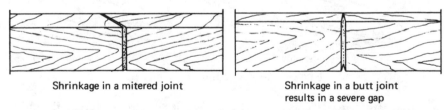

Shrinkage in a mitered joint   Shrinkage in a butt joint
                                      results in a severe gap

Figure 12-4 Shrinkage in Joints, Spliced and Butted

The frieze board is also used on gable ends in a flush cornice. The ridge cut is made along the same lines as a ridge cut of a rafter. A bevel square set to the correct angle is an excellent tool to use to lay out this cut.

The bevel square along with the combination square should be used to make the lower cut of the frieze board (Figure 12-5). The bevel square will provide the angle of the face cut. The combination square will provide the 45-degree (closed) angle needed to complete the layout. When the lower compound miter cut is made, it must fit the horizontal frieze board's mitered end.

To add styling, a molding or moldings can be added to the frieze board in flush cornice. Bed molding (Figure 12-6) and crown molding

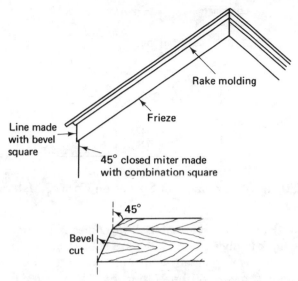

Figure 12-5 Gable-End Compound Miter Cut

Figure 12-6 Moldings on Cornice

are frequently used. These moldings are cut in a miter box. All splices must not only be cut on 45-degree angles but also glued with exterior-type glue for a first-quality job. Since as a rule the moldings are placed along the top edge of the frieze, you should treat them as ceiling molding. Refer to Chapter 9, Routine IT5, Installing Ceiling-Border Molding, for detained instructions on cutting moldings if you are unfamiliar with cutting moldings of the bed and crown variety.

OPEN-RAFTER-END CORNICE

The principal member in the fabrication of *open-rafter-end cornice* is the freize board. The frieze board must be cut around the rafter overhang (Figure 12-7). This means that each freize board must be custom-fitted.

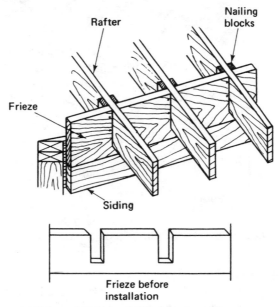

Figure 12-7 Open-End Rafter Cornice

Some provisions for cornice finish must be made before the roof's sheathing is installed. Of primary concern is the need for each rafter to be perpendicular (no twists or leaning). Blocks cut to fit between the rafters and positioned flush with the outside of the wall surface may be installed. Or a block may be nailed with its edge plumb to the outside wall surface. Either method provides a nailing surface for the frieze board.

With some form of blocking/nailing surface installed, the frieze board may now be fitted. Measure and rip the frieze board to the required total width. Rabbet the back side if required. Tack-nail the board against the rafter bottom edges, allowing some end overhang to be cut to fit later. With your ruler as shown in Figure 12-8, mark each side of each rafter on the frieze board. Next measure the total height of the rafter along the plumb line and set the combination square accordingly. Take the frieze board to the sawhorses and make all the lines necessary to define the rafter cutouts needed.

Cut each notch so that the pencil mark remains on the frieze board. Back the frieze board by placing it on a 2 x 6 before chiseling the block of material to be cut away. Chisel the material away, trimming the stock *even* with the line that you drew using the combination square.

Fit the piece in place on the wall. When it is fitted, mark the overhanging end for the miter needed to fit the gable end or the other wall.

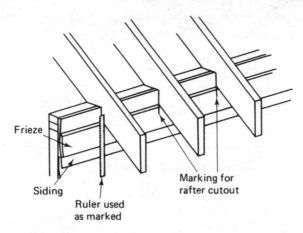

Figure 12-8 Marking Rafters on a Frieze Board

Also, mark the opposite end for an open-cut splice or for a 45-degree closed cut for a corner joint. If a butt joint on an interior corner was needed, the board would have been so placed before marking. Cut the board as required and install it with 8d finishing nails. Set the nails below the wood surface with a nail set and hammer.

Proceed to add molding as required in the same manner as for a flush cornice.

BOXED (CLOSED) CORNICE

The *boxed cornice* usually requires additional framing materials. Figure 12-9 shows the basic elements of a boxed cornice. The structural members needed are usually 2 x 4 lookout ledgers and lookouts. The ledger is installed on the studs or sheathing of the wall on a line level

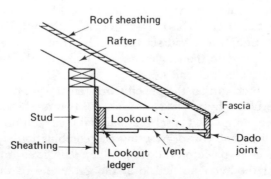

Figure 12-9 Box Cornice Configuration

with the lower edge of the rafters (see Figure 12-9). This member is nailed to the wall with 12d common nails along a previously snapped chalk line.

A measurement is taken from the lookout ledger to the rafter end, and sufficient lookouts are cut so that one can be nailed to each rafter end and toe-nailed to the lookout ledger with 8d common nails.

Before nailing the lookouts, a line must be stretched along the outer edge of the rafters, from end to end. Install the line 3/4 in. above the bottom of the rafter. Check all rafters along the line by using a scrap piece of 1 x 2 stock lumber 6 in. long to detect any rafters that may be hanging too low or not low enough.

Where a rafter is too low, mark it for trimming and cut the excess with a saw. If the rafter is too high, mark the 3/4-line position and, when nailing the lookout, use the 3/4 mark as a reference point.

With the lookouts installed, the frieze board can be fitted and installed as described previously. In this application it is installed from the last course of siding to the *bottom* side of the lookout ledger. Nail it with 8d or 10d finishing nails.

The fascia board is installed next. If a commercial, already grooved fascia board is used, measure from the bottom edge of the lookout to the top of the roof sheathing and rip the board accordingly. If the fascia board is made from standard stock, a dado (groove) must be cut, as shown in Figure 12-10. A power saw with guide may be used, or a router with a dado bit and guide may be used. The groove must be made wide enough to easily insert the soffit board. If using 1/4-in. plywood for soffit material, make the grove 5/16 to 3/8 in. wide. If using 3/8-in. plywood as a soffit, make the groove 7/16 to 1/2 in. wide. Make all grooves 5/16 to 3/8 in. deep. Position the groove so that 1/2 to 5/8 in. of stock is left between the bottom lip of the groove and the edge of the fascia.

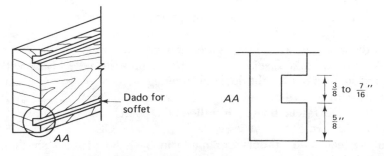

Figure 12-10 Dado in Fascia Board

Precut one end of the fascia board for either a corner joint or a splice. Hold it in position and mark the other end for cutting. Cut as marked.

Install the fascia board so that the end is properly aligned and the top lip of the groove is even with the lower edge of the lookouts. This second part of the task may be done by visually sighting the surfaces to match; it may be done by inserting a guide in the groove and butting the guide to the lookout. Face-nail into the lookout and rafter end with 8d galvanized finishing nails. Set nails with a 1/16-in. nail set and hammer. (*Tip:* If you have built a scaffold for part of the run of a wall and will need to move it to complete the cornice installation, complete as much of the cornice lookouts, frieze, fascia, soffit, and moldings as possible before moving the scaffold.)

Continue installing the fascia as required. On gable ends use the standard rafter ridge cut at the ridge and a compound miter cut at the eave. The fascia will generally be installed on the fly rafter.

Where used with a 12-in. or wider overhang, the boxed cornice requires a return on the gable ends. Figure 12-11 shows the situation and a solution. The lookout ledger should have been extended to the fly rafter's position. A lookout should be nailed to the fly rafter and lookout ledger, and a 2 x 4 filler should be installed. Blocking must be installed to allow nailing the panel on the inside of the rake (Figure 12-11, point 1). A piece of 1 x 12 stock must be fitted over the exposed overhang end. If a design feature such as a cove or double-S curve is used, it must be cut while the piece or pieces are on the ground.

One piece of 1 x 12 stock will usually do the job. Cut it long enough to allow for any design features and some waste. Hold it against the underside of the gable-end fascia at the eave.

Mark the position of the underside of the soffit so that the same amount of lip will be seen below the soffit. Trim the board to the line drawn. Groove for a soffit if required. Again put the piece of 1 x 12 in place and mark the wall line of the overhang plus 1 in. Remove the stock to the bench and draw and cut the design feature. Sand the cut edge after cutting. Try the piece for fit and appearance. If it fits, make a pattern of the piece for use on the other corners.

Prepare some waterproof glue, and glue the edge of the board to be joined to the rake fascia. Nail the board into the rake fascia with 6d finishing nails and into the lookout framing with 8d finishing nails. Set the nails with a 1/16-in. nail set and hammer.

With the fascia boards installed, the soffit may be prepared and installed. As a rule, a soffit is made from 3/8-in.-CD exterior-grade plywood. At least two cutouts are made in each 8-ft length so that the ventilator screens can be installed. Ventilating is essential to good attic

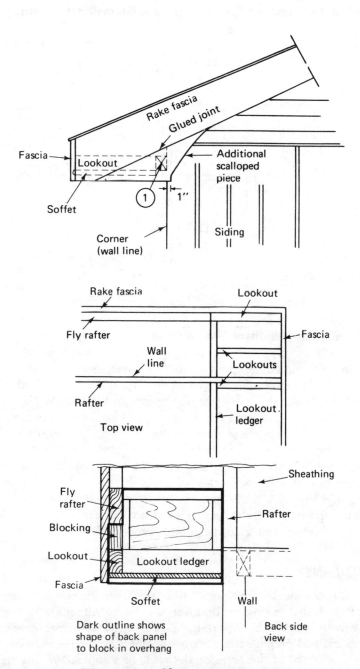

Figure 12-11 Closing In a Return

air flow, heating, and air conditioning. In some installations 1/4-in. plywood of the same quality is used as soffit material (Figure 12-12).

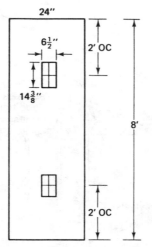

Figure 12-12 Soffit Layout for Ventilation

Measure the width of the soffit area to be covered. Rip the required sheets of plywood accordingly. Lay out and cut two 4 in. x 14-1/4 in. rectangular holes in each full sheet, one in each 4-ft half.

Cut the first piece for length so that its end breaks on the center of a lookout. Insert it in the fascia's groove and nail it to the lookouts with 4d and 6d galvanized finishing nails. Continue installing pieces until the eaves are complete. Use the same process on gable ends.

The only task remaining is to install the moldings. Two areas generally get molding: the roof line of the fascia board and the joint created by the frieze and soffit boards. A piece of 1 x 2 stock, or bed or cove molding in a variety of sizes, may be used for these.

When preparing these moldings, always cut all splices and ends at 45-degree angles and glue the joints before nailing.

ROUTINES

The following six routines outline the methods for installing each piece of stock that is used in building a cornice. Although each job can be done with hand tools, a portable power saw and/or router would be useful to cut and prepare the various members of the cornice. Since this work is performed at the edge of the roof, be reminded of the need for ladders and/or scaffolds and a helper.

BC or 1: INSTALLING FRIEZE BOARD

RESOURCES

Materials:
Plans and specifications with cornice details

_____ lineal feet 1 in. or 5/4 in. x _____ of frieze board.
   no.   select one   width

The *number* of lineal feet is calculated from the plan and equals the total length of all cornice, including gable ends, and the *width* equals the size or sizes of boards needed to cover the area from the last course of siding or brick to the lookout ledger or rafter top edge.

1 lb 8d galvanized finishing nails per 20 lineal ft of cornice

Tools:
1 pair sawhorses
1 no. 8 crosscut handsaw
1 13- to 16-oz claw hammer
1 combination square
1 1/8-in. nail set
1 chalk line
1 ladder (length according to need)
1 level (optional)
1 framing square (optional)
1 set of ladder scaffold jacks and planks (optional)
1 set of wall scaffold jacks and planks (optional)

ESTIMATED MANHOURS

1 to 3 hours per 24 lineal ft

PROCEDURE

Note: Installation steps for specialized cornices are separated and listed under the cornice type: open-end rafter, flush, boxed, or rake.

Step 1 (All) Stretch the chalk line along the last course of siding or brick to verify the straightness of the material. Record irregularities, if any.

Step   2   (All) Measure down from the walls' plate to approximately 1/2 in. below the top of the last course of siding or brick. Snap a chalk line.

Step   3   Lay out the frieze material across the sawhorses and rabbet the bottom inside edge, *if required*.

> Use step 4 for open-end rafter cornice.
> Skip to step 12 for flush rafter cornice.
> Skip to step 15 for boxed cornice.
> Skip to step 17 for rake cornice.

Step   4   Plumb a line with your level along the end rafters where the *inside* surface of the frieze will align.

Step   5   Snap a chalk line (if no sheathing is installed on the roof) along the top of the rafters, connecting the plumb lines marked.

Step   6   Temporarily tack the first (and subsequent) frieze board in place under the rafter overhang.

Step   7   Mark for a cutout of each rafter, and measure the depth of each notch.

Step   8   Remove the frieze to the sawhorses and cut out each notch to a depth equal to the rafter's width.

Step   9   Cut the end of the frieze on a 45-degree angle open to the outside if splicing is needed.

Step 10   Insert the frieze and try for fit. If satisfactory, mark the frieze ends for 45-degree cuts. Cut the angles and install the frieze.

Step 11   Face-nail the board into the studs and plate and toe-nail into the rafters, *keeping* the frieze even with the chalk line along the rafter top. (This completes installation of an open-end rafter frieze.)

Step 12   Measure the total width of frieze needed from the chalk line to the top of the rafter ends.

Step 13   Rip the frieze to the width needed. Cut a 45-degree angle on one end for an outside corner (if required).

Step 14   Position the frieze on the wall and mark for a splice on the opposite end. Cut a 45-degree cut accordingly. Nail in place. (This completes installation of a flush-mounted-cornice frieze.)

Step 15   Measure the width of the frieze needed from the chalk line to the underside of the lookout ledger. Rip the frieze piece(s) accordingly.

Step 16   Cut 45-degree cuts according to the requirements for a splice and returns. Nail in place. (This completes installation of a boxed-cornice frieze.)

Step 17 Measure the width of rake frieze board needed at the point where the horizontal overhang frieze board is cut for a return. Rip the required stock.

Step 18 Make a ridge cut on one end. Mark the lower point or make a splice cut, as required.

Step 19 Miter a 45-degree cut along the line described in step 17. Nail in place. (This completes installation of a rake-cornice frieze.)

BC or 2: INSTALLING LOOKOUTS AND LOOKOUT LEDGERS

RESOURCES

Materials:
_____ lineal ft 2 x 4. One piece is needed for each lookout and no.

enough is needed for the ledger. Use your plan.

1 lb 12d common nails per 4 lineal ft of overhang
1 lb 8d common nails per 6 ft of lookout ledger

Tools:
1 no. 8 crosscut handsaw
1 16-oz hammer
1 24-in. level
1 miter square
1 framing square
1 combination square
1 chalk line
1 pair sawhorses
Ladders and scaffold jacks (optional)

ESTIMATED MANHOURS

1 hour per 4 ft of overhang

PROCEDURE

Preliminary step   Verify with a chalk line that all outer edges *and* lower cuts on the rafters are equal. Trim if required.

Step 1   With the level, mark several points along the wall level with the bottom edge of the rafter.

Step 2   Snap a chalk line that connects the points.

Step 3   Position a lookout ledger to break on the center of a stud and nail in place along the chalk line (2 x 4 above the line) with 12d nails.

Step 4   Complete the ledger installation.

Step 5   Precut 2 x 4 lookouts.

Step 6   Nail the lookouts to the rafter ends with three 12d nails and toe-nail to the lookout ledger with four 8d nails.

BC or 3: INSTALLING FASCIA

RESOURCES

Materials:
_____ lineal ft of 1 x _____ fascia board. The lineal footage
    no.   width

is equal to the total length of overhang (use your plan) *plus* 10 percent for waste.

1 lb 8d or 10d (select one) galvanized finishing nails for each 6 ft of fascia

Tools:
1 no. 8 crosscut handsaw
1 16-oz claw hammer
1 combination square
1 1/16-in. nail set
1 pair sawhorses
1 pair scaffold jacks and ladders (optional)
1 table saw or portable power saw if dado is cut on the job (optional)

ESTIMATED MANHOURS

For two people   30 minutes per 12 ft

PROCEDURE

Preliminary step   Perform Routines BCor1 and BCor2 before proceeding with this routine.

Step 1   Prepare stock for fascia use.
a. Trim to proper width.
b. Dado for soffit (if required).
c. Make a 45-degree cut if splicing is used.

Step 2   Install the fascia with the splice cut (if used) breaking on a rafter and the dado's top edge *flush* with the bottom edge of the lookouts. Nail with two to three nails per rafter.

Step 3   Trim the overhanging fascia flush with the roof end line.

Step 4   Repeat steps 1 through 3 as required.
Note: Adjust the loose end of the fascia up or down as the nailing progresses to take out any bow in the fascia board.

Step 5   For rakes and gable ends, precut the gable-end rafter cut (and splice cut if required). Nail in place. Trim the excess fascia board with a handsaw along the lower roof line.

BC or 4: INSTALLING SOFFIT

RESOURCES

Materials:
_____ 4 x 8 x 1/4 in. or 3/8 in. exterior plywood grade CD or better.
   no.   select one
The number of strips made from one sheet of plywood, _____,
divided by the total lineal feet of soffit, _____, equals the number of sheets needed.
1 lb 4d galvanized finishing nails per two sheets of plywood
2 screen vents per sheet

Tools:

1 no. 8 crosscut handsaw

1 pair sawhorses

1 chalk line

1 framing square

1 3/4-in. wood chisel

1 13-oz hammer

1 1/16-in. nail set

ESTIMATED MANHOURS

For two people   1 hour per sheet

PROCEDURE

Step 1   Precut plywood into strips 1/4 to 3/8 in. less than the total width of the soffit area and 8 ft long.

Step 2   Center and cut out two areas for screen vents.

Step 3   Measure and cut each sheet for length before installing. Break the joint on a lookout.

Step 4   Insert the soffit piece into the fascia dado, raise to the lookouts, and nail at 5- to 6-in. intervals along each lookout. Use 12-in. spacing along the lookout ledger, 3 to 4 in. at each joint of plywood.

Step 5   Repeat steps 3 through 4 until all the pieces are installed.

BC or 5: INSTALLING CORNICE MOLDING

RESOURCES

Materials:

_____ lineal ft of molding type _____ and size _____
     no.

1 lb 8d galvanized finishing nails per 50 ft of molding

Tools:
1 miter box
1 no. 8 to no. 10 crosscut handsaw or straight-back saw
1 13-oz claw hammer
1 1/16-in. nail set
1 ladder
1 pair sawhorses

ESTIMATED MANHOURS

1 hour per 50 ft

PROCEDURE

Step 1 Precut a 45-degree corner on one end of the molding if starting from a corner or splicing. Insert the molding upside down into the miter box.

Step 2 Hold the molding in place and tack-nail. Mark the opposite end for cutting. Take the molding to the sawhorses.

Step 3 Insert the molding upside down in the miter box and cut the required 45-degree angle.

Step 4 Nail in place with 8d nails at 16-in. intervals. (Keep the back surfaces flat on frieze and soffit or fascia.)

Step 5 For gable-end molding, make a compound miter cut for rakes and gable ends.
a. Mark the molding for a cut at the ridge, and cut.
b. Mark the lower end of the molding for a ridge cut *and* 45-degree cuts.
Note: Cut both pieces freehand or use a miter box.

BC or 6: INSTALLING CORNICE RETURNS

RESOURCES

Materials:
1 in. x _____ x _____ of frieze and/or fascia as derived
        width   lineal ft

from plans

_____ lineal ft _____ molding
    no.   type

1/2 lb 8d galvanized finishing nails per corner

1/4 lb 6d galvanized finishing nails per corner

_____ lineal ft of rain cap
    no.

1 tube caulking compound

Tools:

1 pair sawhorses

1 no. 8 crosscut handsaw

1 13-oz claw hammer

1 combination square

1 framing square

1 miter square

1 coping or keyhole saw

1 nail set

1 power saber saw (optional)

1 stepladder or extension ladder (optional)

ESTIMATED MANHOURS

2 hours per corner

PROCEDURE A: BOXED CORNICE RETURN

Preliminary step   Install horizontal cornices and rakes/gable-end cornices.

Step 1   Measure and cut the return material for length.

Step 2   Temporarily install the material in place and mark for a bevel cut to fit the rake.

Step 3   Remove, cut, and verify fit. Install.

Step 4   Lay out the scallop or cove piece for return if required. Unless matching other returns, make the design freehand or with dividers.

Step 5   Cut and sand the scallop or cove cut.

Step 6   Lay out the remaining pieces before fastening the piece to the rake and fascia returns.

Step 7   Nail in place and set all nails. Caulk the joints.

PROCEDURE B: FLUSH CORNICE RETURN

Preliminary step Complete the horizontal frieze-board installation but not the gable.

Step 1 Cut the frieze board for the return length desired.

Step 2 Fit the piece cut to join the frieze board. Nail in place (preferably over siding).

Step 3 Cut and fit the molding on the return's upper edge (to match the molding on the frieze). Nail in place.

Step 4 Cut and install the rain drip cap flush with the siding on the *top* of the return. Caulk all joints where water might enter.

Step 5 Cut and install the lower piece of gable end frieze board to fit on the rain cap.

13

WINDOW-UNIT INSTALLATION

BASIC TERMS

Blind stop outside stop that retains sash.

Casing outside trim nailed to blind stop.

Drip cap molding installed on casing at the top of the window

Jamb sides (styles) and head of a window frame.

Mullion connecting link (member) between two standard window units.

Muntins small dividers in a sash used for colonial- and western-style windows.

Parting strip stop between inner and outer sash on a double-hung window.

Sash window subassembly, containing glass.

Sill base of a window frame.

Window units are installed concurrently with the application of sheathing, immediately after the sheathing is installed, or after the exterior siding is installed. The type of window unit and the type of wall covering combine to dictate when and how the window will be installed.

SCOPE OF THE WORK

Installing a window unit is a relatively simple task, but two people are generally required, even with lightweight aluminum windows. Windows are installed from the outside of the building. Therefore, a ladder or scaffold will be needed to get to the window area, and later to nail the unit in place. The second person, working from the inside, can adjust the unit's position for centering, plumb, and level while the person outside nails the unit in place.

The following paragraphs provide background for the most common types of window units: *double-hung, casement,* and *awning.* Each type is available in both wood and aluminum.

DOUBLE-HUNG WINDOWS

The double-hung window is perhaps the most familiar type. It consists of an upper and a lower sash that slide vertically in separate grooves in the side jambs or in full-width metal weatherstripping (Figure 13-1). This type of window provides a maximum face opening for ventilation—one-half the total window area. Each sash is provided with springs, balances, or *compression weatherstripping* to hold it in place in any location. Compression weatherstripping, for example, prevents air infiltration, provides tension, and acts as a counterbalance. Several types are manufactured which allow the sash to be removed for easy painting or repair.

The *jambs* (sides and top of the frames) are made of nominal 1-in. lumber. Their width is calculated for use with drywall or plastered interior finish. Sills are made from nominal 2-in. lumber and are sloped at about 3 in 12 for good drainage (Figure 13-1B). Sash are normally 1-3/8 in. thick and wooden combination storm and screen windows are usually 1-1/8 in. thick.

Sash may be divided into a number of lights by small wooden

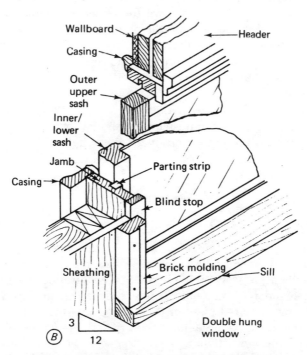

Figure 13-1 Double-Hung Window Units

members called *muntins*. A ranch-style house may look best with top and bottom sash divided into two horizontal lights. A house in the colonial or Cape Cod style usually has each sash divided into six or eight lights. Some manufacturers provide preassembled dividers that snap in place over a single light, dividing it into six or eight lights. This simplifies painting and other maintenance.

Assembled frames are placed in the rough opening over strips of building paper that have been nailed around the perimeter to minimize air infiltration. The frame is plumbed and nailed to the side studs and header through the casings or the blind stops at the sides. Where nails are exposed, such as on the casing, galvanized nails should be used.

Hardware for double-hung windows includes the sash lifts fastened to the bottom rail, although they are sometimes eliminated by providing a finger groove in the rail. Other hardware consists of sash locks or fasteners located at the meeting rail. They not only lock the window, but draw the sash together to provide a "windtight" fit.

Double-hung windows can be arranged in a number of ways: as a single unit, doubled (or mullion) type, or in groups of three or more. One or two double-hung windows on each side of a large stationary insulated window are often used to give the effect of a window wall.

If a wide blind stop is used on the outside of the window, the siding will be installed on top of it (Figure 13-2). If a narrow blind stop is used, the brick molding will cover the siding *or* the siding will butt to the brick molding (Figure 13-2B). The type of unit purchased will dictate the installation method used.

Aluminum windows are manufactured slightly differently from wooden windows. The aluminum window shown in Figure 13-3 has the customary flange on the outer portion of the window. Nails are driven through the flange into the sheathing and studding. Siding is butted to the frame styles and cut along the sill and head.

CASEMENT WINDOWS

A casement window consists of a side-hinged sash, usually designed to swing outward (Figure 13-4) because this type can be made more weathertight than the in-swinging style. Screens are located inside these out-swinging windows and winter protection is obtained with a storm sash or by using insulated glass in the sash. One advantage of the casement window over the double-hung type is that the entire window area can be opened for ventilation.

Weatherstripping is also provided for this type of window. The units are usually received from the factory entirely assembled, with hardware in place. Closing hardware consists of a rotary operator and sash

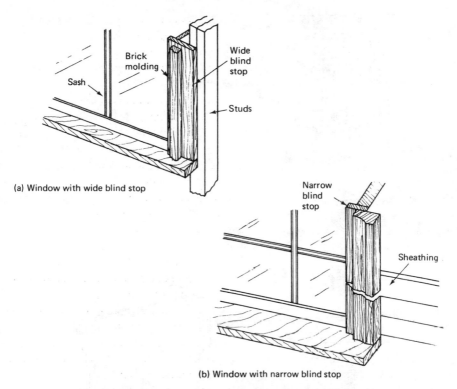

(a) Window with wide blind stop

(b) Window with narrow blind stop

Figure 13-2 Window with Wide Blind Stop

Figure 13-3 Aluminum Double-Hung Window Unit

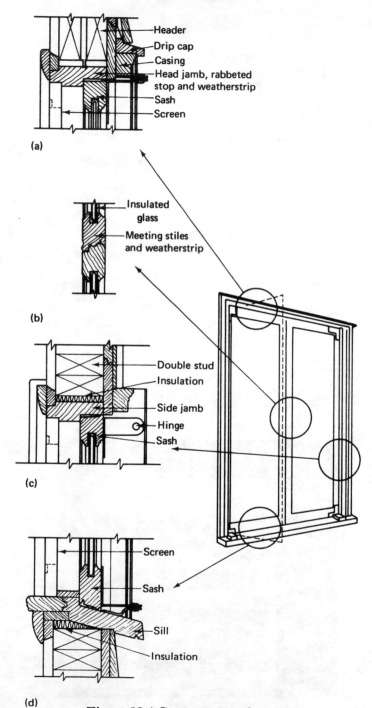

(a)

(b)

(c)

(d)

Figure 13-4 Casement Window Unit

lock. As in the double-hung units, casement sash can be used as a single unit, as a pair, or in combinations of two or more pairs. Style variations are achieved by divided lights.

Aluminum casement windows are similar to the wooden ones. A flange is provided for installation purposes.

AWNING WINDOWS

An awning window unit consists of a frame in which one or more operative sash is installed (Figure 13-5). They often are made up for a large window wall and consist of three or more units in width and height.

Sash of the awning type are made to swing outward at the bottom. A similar unit, called the *hopper* type, is one in which the top of the sash swings inward. Both types provide protection from rain when open.

Jambs are usually 1 and 1/16 in. or thicker because they are rabbeted, while the sill is at least 1 and 5/16 in. thick when two or more sash are used in a complete frame. Each sash may also be provided with an individual frame so that any combination of width and height can be used. Awning window units may consist of a combination of one or more fixed sash, with the remainder being the operable type. Operable sash are provided with hinges, pivots, and sash-supporting arms.

If aluminum windows are used, a basic wooden frame unit can be made to order from a local mill or you can design and build one from no. 1 clear 2 x 6 fir or white pine.

INSTALLATION OF WINDOW UNITS

Only two methods of installing windows exist in wooden frame wall construction: (1) the unit is installed over the sheathing and/or sheathing and siding, or (2) the window is installed on the studding. The second type is seldom used today.

We shall call installing a window unit on the sheathing and/or siding type A (Figure 13-6). This type includes wooden frame window units and almost all aluminum windows. The window unit is installed from the outside. It is inserted in the opening, shoved to the header and lowered slightly, and centered before nailing. The reason for lowering the window away from the header is to allow for leveling and plumbing.

Nails are used to hold the window unit in place. First, a nail is installed near the top of one side of the window. Next, the window is leveled and nailed at the opposite side near the top. Follow this by plumb-

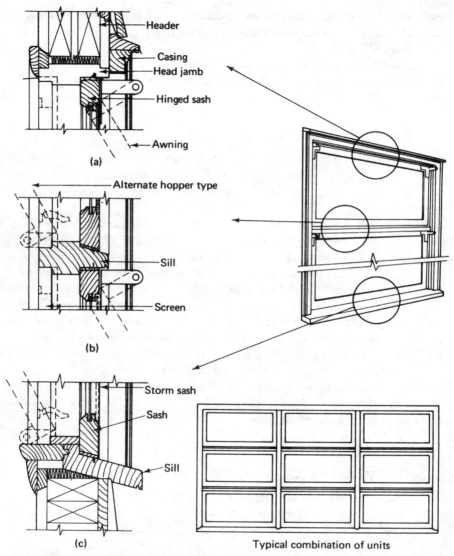

(a)

(b)

(c)

Typical combination of units

Figure 13-5 Awning Window Unit

ing the style and nailing it. Finally, nail the entire outer perimeter of the window at 6-in. intervals.

We shall call the second type of wooden window unit, one with a wide blind stop, type B (Figure 13-7). This older type of window requires that its installation be made directly onto the stud window framing. If this type is used and the sheathing is installed prior to window installation (the usual practice), keep the sheathing back far enough to

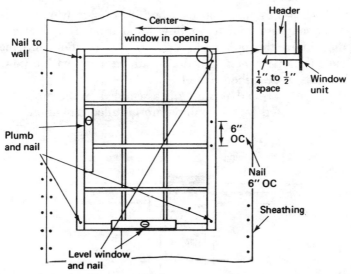

Figure 13-6 Installing a Window Unit with Narrow Blind Stop and/or an Aluminum Window

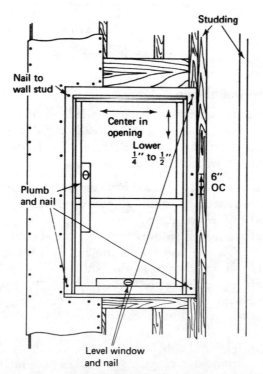

Figure 13-7 Installing a Window with Wide Blind Stop

allow the blind stop to pass. You may find it easier to trim the sheathing with a power saw just prior to window installation. If you elect to trim the opening, insert the unit to its approximate position and mark the cutting line on the perimeter. Cut the sheathing away with a power saw. Install the unit as described earlier.

If a mullion is needed, it is usually assembled on the ground (Figure 13-8). All units united by the mullion are installed simultaneously.

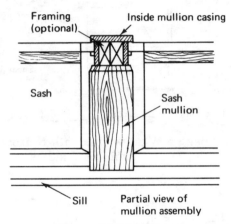

Figure 13-8 Mullions

Testing the Installation

After the window units have been installed, all hardware should be adjusted to operate smoothly and to be weathertight when the sash is closed and locked. Hardware and parts should be lubricated as necessary. Adjustments and tests should be as follows.

1. Double-hung windows should have the balances adjusted to allow the sash to remain steady in all positions, and the guides should be waxed or lubricated.
2. Casements equipped with rotary operators should be adjusted so that the top of the sash makes contact with the frame approximately 1/4 in. in advance of the bottom.
3. Casements equipped with friction hinges, or friction holders, should be adjusted so that they latch securely.
4. Projected sash should have arms or slides lubricated and adjusted to close with a slight tension on the arm.
5. Awning windows should have the arms to the sash adjusted so that the bottom edge of each sash makes continuous initial contact with frames when closed.

6. Where windows are weatherstripped, the weatherstripping should make weathertight contact with the frames when the sash is closed and locked. The weatherstripping should not cause binding of the sash or prevent closing and locking of the windows.

ROUTINES

There are only two routines provided for window installation. Routine BW1 will be used more often because it outlines instructions for modern windows. Routine BW2 is provided for windows that have a wide blind stop. In both routines the material requirements require you to compile a list of all the windows needed for the project. State each by size and type. In each routine two people will be needed to install a window.

BW1: INSTALLING ALUMINUM AND WOODEN TYPE A WINDOW UNITS

RESOURCES

Materials:
List of windows by size taken from plan or specifications
Average 12 6d common nails per window (aluminum)
Average 12 10d finishing or casing nails per window (Wooden type A)

Tools:
1 13-oz claw hammer
1 24-in. level
1 sawhorse
1 ladder
1 scaffold

ESTIMATED MANHOURS

For two people   15 to 20 minutes per window

PROCEDURE

Step 1 Insert the window unit into the opening from the outside.

Step 2 The person inside pulls the window in to the wall and up to the header. Lower 1/4 to 1/2 in.

Step 3 The person outside nails one outside corner with 6d common or 10d finishing nails.

Step 4 Place the level on the window sill and adjust for level. Nail the opposite upper corner of the window.

Step 5 The person inside places the level in a plumb fashion on the window style (the vertical structural member) and adjusts for plumb; the person outside nails.

Step 6 Nail all surfaces at 6-in.-OC spacings.

BW2: INSTALLING WOODEN TYPE B WINDOW UNITS

RESOURCES

Materials:
List of windows by size taken from plan or specifications
Average 12 8d common nails per window

Tools:
1 13-oz claw hammer
1 24-in. level
1 no. 8 crosscut handsaw
1 sawhorse
1 ladder
1 scaffold

ESTIMATED MANHOURS

For two people   15 to 20 minutes per window

PROCEDURE

Step 1   Insert the window until into the opening from the outside.

Step 2   The person inside pulls the window in to the wall and up to the header. Lower 1/4 to 1/2 in.

Step 3   The person outside nails one outside corner with 8d common nails.

Step 4   Place the level on the window sill and adjust for level. Nail the opposite upper corner of the window.

Step 5   The person inside places the level in a plumb fashion on the window style (the vertical structural member) and adjusts for plumb; the person outside nails.

Step 6   Nail all surfaces at 6-in.-OC spacings.

14

DOOR-UNIT INSTALLATION

BASIC TERMS

Facing/casing trim stock on the outside and/or inside fastened to the jamb.

Jamb side and top structural members of a door unit; parts to which the hinges are fastened.

Sill bottom piece of nominally 2-in. stock on an exterior door unit.

Stop small 3/8 in. x 1 in. x 1/4 in. to 1/2 in. x 1 in. x 5/8 in. stock which seals the door and stops door travel.

A major breakthrough for the beginning carpenter and homeowner has been the manufacturing of assembled door units. Exterior-door-frame units with and without doors have been available for years. Only recently, however, has the interior door unit been developed. Through this technology, installing door units has been changed from a time-consuming and exacting task to a relatively simple task that can be accomplished by persons of little skill.

SCOPE OF THE WORK

There are two aspects to this work: one, the installation of exterior door units; the other, the installation of interior door units. The exterior types require attention to a variety of details before and during installation. Some of these are type of floor, to use or not to use the sill, backing with waterproof paper, nailing and blocking techniques, and trimming the jamb for proper width. The interior jambs, because of their construction, are simple to install and in every case the door is used as a guide in the installation.

One man can usually handle the installation. As a rule, one works on the outside of the exterior door unit and on the hinge side/opening of the interior door unit. Because the trim is nailed to the wall, the door can be shimmed without much trouble. The installation of interior door units is usually completed with the sliding in place of the other half of the jamb and nailing. The lock may be easily installed in the precut interior door and may already be installed. The table in BD1 Routine provides data for the sizes of standard door units and required rough openings.

Figure 14-1 shows an interior and exterior door frame unit. Note that there are both similarities and differences. The similarities are that both frames have *jambs* and *stops*. The differences are that the exterior door frame *may* have a *sill,* and has trim only on the outside, whereas the interior door frame has *trim* on both sides of the jamb. Note also that there is one other difference between the jambs: the exterior jamb is made from one solid piece, whereas the interior jamb is made from two custom-designed pieces of stock.

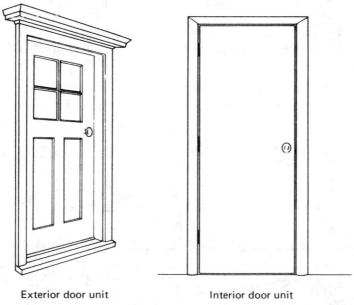

Exterior door unit   Interior door unit

Figure 14-1 Door units

INTERIOR DOOR UNITS

Figure 14-2 shows a better view of the division of pieces of the interior door unit. The two halves of the jamb are separately shaped on large machines to the specifications indicated in the figure. By using this organization the unit may be used on a variety of walls, from paneled to plastered. As indicated, the casing (trim) design is optional, and it is usually stapled, but may be nailed, to the jamb.

Occasionally you will encounter a three-piece adjustable jamb unit (Figure 14-3). The *stop* is a separate piece and may either be stapled

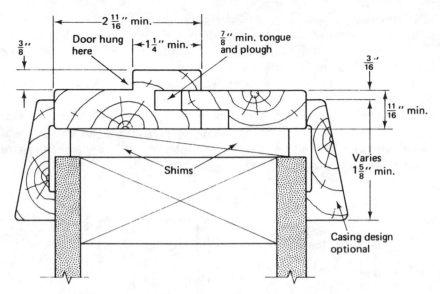

Figure 14-2 Two-Piece Adjustable Jamb Unit

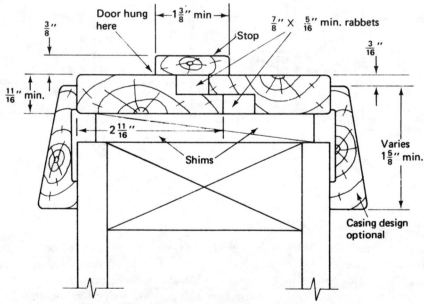

Figure 14-3 Three-Piece Adjustable Jamb Unit

to one side of the jamb or tacked to allow you to position it during installation of the jamb.

The installation of these types of door units follows a specific but relatively easy process. The door is left hanging on its hinges, but the two halves of the jamb must be separated. Nails, which are usually particularly driven through the jamb into the door on the sides and top, must be removed before the jamb can be separated.

The half of the jamb with the door is installed first. The best way to insert it in the opening is to grasp it by the casing and tilt the head inward. Once in the opening shift the unit left and right to center it. While holding it, tack-nail the casing near the bottom hinge. Adjust the jamb left or right as shown in Figure 14-4 until a 1/8-in. space is even across the head; then nail the hinge side casing near the top. Position the

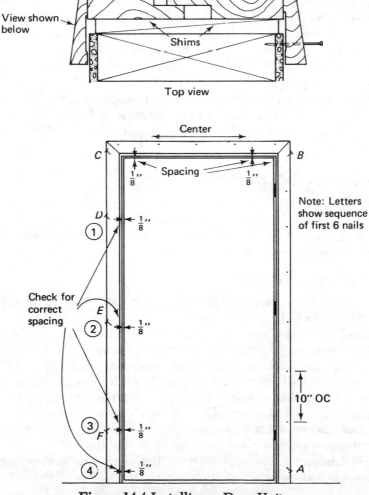

Figure 14-4 Installing a Door Unit

leading-edge jamb for a 1/8-in. spacing between the door and the jamb and nail through the casing.

Cut and insert shims between the jamb and the studs so that they fit firmly but do not force the jamb inward. These shims should be installed near each hinge and lock area. They are held in place by face-nailing through the jamb.

Once the shims are in place and nailed, try the door again for proper closing. If it closes properly, proceed with the installation of the remaining half of the jamb. Slide it into the grooved area and, when the casing makes contact with the wall, nail it. The nails around the casing should be placed on a level with those nailing the casing to the jamb. Set the nails with a 1/16-in. nail set.

Tip: This work is trim and so is finish work. Therefore, any hammer mark will be seen clearly even after painting. To preclude making these marks, refrain from driving the nail *home* with the hammer. Instead, use the nail set to drive home and set the nail. In addition, by using a 13-oz hammer instead of a 16-oz hammer, any slips and hammer marks will be lighter and less noticeable.

With the unit installed, the lock or passage set may be installed. The door is precut and drilled for the cylinder lock. Follow the directions given in the box containing the lock and refer to Chapter 15 if necessary.

EXTERIOR DOOR UNITS

Recall that the sill is made from 2-in. stock and is dadoed into the jamb sides. In certain applications, the sill may be used; in others, it will not. These will be treated separately because different subordinate tasks are required to complete the installation.

When Using the Sill

As a rule the door sill is left intact when the unit is installed over a wood-framed floor. To properly set the unit in place requires some calculation and planning. From your plan, extract the finished height of the floor. The cross-section detail should provide this figure. If it is not available, it might be best to draw the detail now. Notice in Figure 14-5 that we have selected a finished floor height of 3/4 in. *above* subfloor height. Also note that the subfloor is 3/4 in. thick. Therefore, the total thickness of both floors is 1-1/2 inches. The sill, as already stated, is made from 2-in. stock (actual 1-1/2 in.) and, because it is on a slope of 3 in 12, its interior rise (edge surface) will measure 1-9/16 in.

Obviously, we cannot place it on top of the subfloor, so the subfloor must be cut away. In addition, we must cut some of the joist fram-

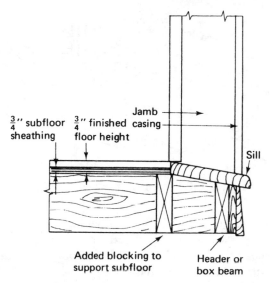

Figure 14-5 Planning a Door-Sill Height

ing to account for the sill's slope, in order to have the inside top edge of the sill measure 3/4 in. above the subfloor.

A simple method of preparing the sill area is as follows. Measure a like distance from the outside surface of the wall inward on both sides of the opening equal to the width of the jamb. Draw a connecting line. Remove all the nails from the sheathing and cut away the stock. Set the miter square for 3 in 12 with a framing square or by using the sill/jamb as a guide. Hold the handle flush with the box beam (joist) and measure the space between the blade and the beam top. This is the amount of joist material that must be cut away from the outer edge of the box beam. Notch several places across the beam to the line drawn and chip away the stock. Slope all joists (if any) from the outer edge to nothing at the subfloor line (Figure 14-6).

Since we cut away some of the subflooring, it has no support. This area will be heavily traveled, so 2 x 4 blocking must be installed. Cut and install 2 x 4 blocking tightly against the underside of the subfloor. If the blocks must be nailed in the exposed area, chip away the tops of them to conform to the slope. Nail the subfloor securely to the blocks.

Insert the door unit, minus the door, in the opening, and tack-nail it to hold it. Review your preparations and make sure that the top of the sill is at the proper height (3/4 in. in our example). Remove the unit, make the necessary adjustments, and install building paper around the door opening.

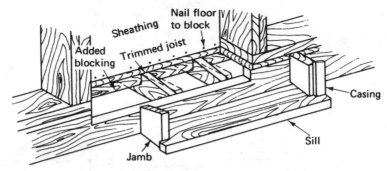

Figure 14-6 Preparing for Sill Support

Reinsert the frame, center it in the opening, and tack-nail it at the bottom sides of the casing. Install the door, adjust the frame left and right until the door fits well, and face-nail it through the jamb and shims into the studs with 10d finishing nails.

When Not Using the Sill

As a rule the wood sill is removed from an exterior door unit when the unit is installed over a cement floor. Because the sill is removed, some preliminary preparation must be performed.

Lay the door unit across two sawhorses *after* removing the door. Study for a moment the way the sill is dadoed into the side jamb. Figure 14-7 shows a typical dado installation. Probably the easiest way to disassemble the unit is to use a piece of stock 12 in. long to drive the jamb away from the sill. Always strike the 2 x 4, *never* the jamb directly, with the hammer. When the jamb and sill are separated about 1 in., try to close the pieces again by taping the jamb's outer surface for the purpose of freeing the nails. If the nails break free, use your hammer to extract them. If they don't break free, you may chisel away some stock around the nailheads or drive a small crowbar around the nailheads.

After the sill has been removed, a guide *must* be installed across the lower portion of the jamb (Figure 14-8). Mark the length of the guide by holding one end of a 1 x 2 or 1 x 3 against the inside of the *casing* at the head of the jamb, and make a mark even with the inside of the opposite jamb. Cut the stock along the mark. Next measure a point 12 in. from the bottom on each side of the jamb. Nail the separator guide tightly to the casing and even with the 12-in. marks.

With the sill removed, some excess jamb stock must be trimmed away. As a rule, you should cut the jamb even with the *top* of the dado where the dado enters between the door stop and the hinge area. Refer

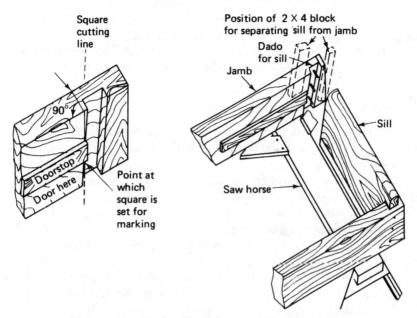

Figure 14-7 Dadoed Jamb and Sill Construction

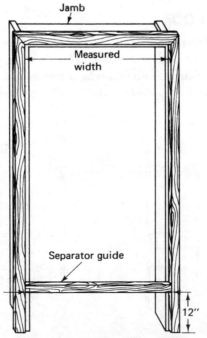

Figure 14-8 Installing the Separator Guide

again to Figure 14-7 and notice where the combination square is to be placed. The cut will always be made at 90 degrees.

With the preliminary preparation completed, the installation of the jamb is the same as that for the other types of jambs.

SHIMS AND SHIMMING

Shims may be made of stock lumber cut in various thicknesses from 1/8 to 3/4 in., or they may be wedge-shaped. Cedar shingles may also be used as shims.

If wedges are used for shimming, one piece must be installed from each direction, as shown in Figure 14-2. The advantage of the wedge is its ability to expand to a variety of widths. The disadvantage is that there may be some difficulty in installing the wedges.

Stock lumber strips can also be used easily and effectively. It is usually handy to have a variety of thicknesses on hand. Insert enough pieces to firmly fill the space between the jamb and the stud. Nail the shims by driving nails through the face of the jambs, through the shims and into the stud. After nailing, use your crosscut handsaw to cut the excess shim material flush with the edge of the jamb.

SWING OF DOOR

Figure 14-9 shows what *swing of door* represents. When ordering doors you must state whether a right-hand door or a left-hand door is wanted. A *right-hand* door is one in which the latch is on the right when a person faces a closed door on the hinge side. A *left-hand door* is one in

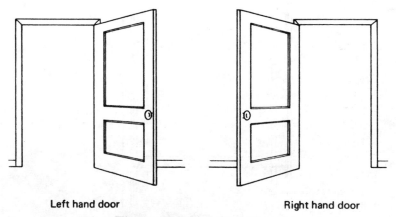

Left hand door   Right hand door

Figure 14-9 Swing of Door

which the latch is on the left when a person faces a closed door on the hinge side.

ROUTINES

Two routines are provided. Routine BD1, Exterior-Door-Unit Installation, is subdivided into two procedural sequences. One sequence is for installing the unit over a wooden joint and floor. The other sequence is for installation over a cement floor. Routine BD2, Interior-Door-Unit Installation, is designed around the standard factory-made two-piece and three-piece door units.

BD1: INSTALLING AN EXTERIOR DOOR UNIT

RESOURCES

Materials:
Door unit _____
               size/type
Blocking material
  wood shingles _____
  wedges _____
  stock _____
1 lb 10d finishing or casing nails
7 lineal ft 30-lb felt (tarpaper)

Tools:
1 no. 8 crosscut handsaw
1 24-in. level
1 6-ft straightedge
1 13-oz claw hammer
1 1/16-in. nail set
1 set wood chisels
1 combination square
1 hand ax (optional)
1 star drill or electric drill and carbide drill bit (procedure A)

3 or more wood/plastic plugs (procedure A)
1 aluminum threshold (procedure A)

ESTIMATED MANHOURS

2.5 to 4 hours

PROCEDURE A: OVER A CONCRETE FLOOR

Step 1 Nail a brace across the outside of the door frame 12 in. up from the bottom.

Step 2 Remove the door from the frame by extracting the pins from the hinges.

Step 3 Remove the threshold from the door frame.

Step 4 Square the bottom of the jamb with the combination square and cut with the saw.

Step 5 Cut the tarpaper into three 12-in. strips; fold the strip in half lengthwise and nail the creased edge flush with the wall door, opening the sides only.

Step 6 Insert the door frame in the opening and center between the studs.

Step 7 Tack-nail the face of the door frame on the singe side at the head.

Step 8 Check the head for level. If level, proceed to step 9; if not, do steps 14 through 17, then return to step 9.

Step 9 Plumb the jamb previously tack-nailed by using the 24-in. level and straightedge. When plumb, tack-nail the bottom face molding.

Step 10 Insert wedges/spacers between the jamb and the stud to fill the space *firmly*; face-nail through the jamb.

Step 11 Install the door on its hinges

Step 12 Close the door and adjust the jamb in or out with spacers until a 1/8-in. space between the door and the jamb is achieved along both the door's leading edge and top.

Step 13 Tack-nail the frame and try the door.

Step 14 Face-nail through the jamb and spacers into the studs after adjustments have been made. Set all nails tacked on the outside trim and install more nails at 8- to 10-in. intervals. (See Routine BDM5 for threshold installation.)

Step 15 Adjust the door jamb up until its head is level, as seen from the level's indication; or

Step 16 Adjust the level up or down to determine the amount of distance out of level in 24 in.

Step 17 Remove the door frame from the opening and cut a corresponding amount of material off the long side (as measured in step 14 or 15).

Step 18 Reinsert the frame and test for a level head. Tack-nail as before. (Return to step 9.)

PROCEDURE B: OVER A WOODEN JOIST AND FLOOR

Step 1 Insert the unit in the opening and tack-nail the frame.

Step 2 Mark the subflooring for trimming along the inside edge of the sill of the door unit.

TABLE BD1: Door Sizes

STANDARD: Width x height x thickness x type, swing

Interior
  2 ft 0 in. x 6 ft 8 in. x 1-3/8 in. interior, LH/RH
  2 ft 4 in. x 6 ft 8 in. x 1-3/8 in. interior, LH/RH
  2 ft 6 in. x 6 ft 8 in. x 1-3/8 in. interior, LH/RH
  2 ft 8 in. x 6 ft 8 in. x 1-3/8 in. interior, LH/RH
  3 ft 0 in. x 6 ft 8 in. x 1-3/8 in. exterior, LH/RH

Exterior
  2 ft 6 in. x 6 ft 8 in. x 1-5/8 in. exterior, LH/RH
  2 ft 8 in. x 6 ft 8 in. x 1-5/8 in. exterior, LH/RH
  3 ft 0 in. x 6 ft 8 in. x 1-3/4 in. exterior, LH/RH
  3 ft 6 in x 6 ft 8 in. x 1-3/4 in. exterior, LH/RH
  4 ft 0 in. x 6 ft 8 in. x 1-7/8 in. exterior, LH/RH

NONSTANDARD
  Standard widths, 7 ft 0 in. height
  Standard widths, 8 ft 0 in. height

DOUBLE DOORS (2 standard/nonstandard doors in one jamb)
  4 ft 6 in. x 6 ft 8 in. x 1-5/8 in.
  5 ft 0 in. x 6 ft 8 in. x 1-5/8 in.
  6 ft 0 in. x 6 ft 8 in. x 1-3/4 in.
  6 ft 0 in. x 7 ft 0 in. x 1-3/4 in.

Step 3 Measure the height of the sill above the subfloor. Record the height.

Step 4 From your plans, determine the height of the finished floor and subtract this height from the sill height measured in step 3. The remainder is the amount of depth needed to be cut from the subfloor and the joists under the subfloor.

Step 5 Remove the door unit and cut the excess material away from the subfloor and joists.

Step 6 Reinforce the sill area by blocking the subfloor under it with 2 x 4.

Step 7 Reinsert the door unit using steps 3 through 12 of procedure A.

BD2: INSTALLING AN INTERIOR DOOR UNIT

RESOURCES

Materials:
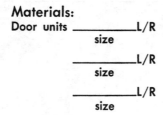
Door units _____L/R
          size

         _____L/R
          size

         _____L/R
          size

24 8d finishing nails per door
Shim material

Tools:
1 13-oz claw hammer
1 no. 8 or no. 10 crosscut handsaw
1 6-ft straightedge
1 6-ft folding ruler
1 1/16-in. nail set

ESTIMATED MANHOURS

30 minutes per door unit

PROCEDURE

Step 1 Remove from the jamb the retaining nails that hold the door front to the jamb and separate the halves of the jamb.

Step 2 Insert the jamb and door into the opening. Center within the opening.

Step 3 With the door closed, tack a nail through the trim at the lower hinge side.

Step 4 Adjust the upper part of the frame until a 1/8-in. clearance is measured across the entire top of the door to the frame. Nail the trim to the wall.

Step 5 Adjust the front frame to the door edge for 1/8-in. clearance and nail to the wall as each adjustment is made.

Step 6 Cut and insert wedge/shims between the jamb and the stud. Face-nail through the jamb into the shim and stud (behind the hinges and lock).

Step 7 Install the other half of the jamb and nail the trim to the wall.

Step 8 Nail all trim pieces at 10-in. intervals and set all nails.

Step 9 Install the lock and knobs in the holes prepared. Install the striking plate in the door jamb.

15

DOOR INSTALLATION AND MAINTENANCE

Felt paper building paper, tar-impregnated, 30 lb weight, to be used as a sealant between threshold and sill.

Head top part of a door or the horizontal member of a door-jamb unit.

Hinge metal assembly, usually containing a removable pin.

Lock set assembly that allows for opening and closing (passage set) a door, or securing (lock and passage set) a door.

Striking plate metal plate installed on the jamb, into which the bolt or plunger fits.

Template cardboard layout of the positions for centering the cylinder and handle assemblies of a lock set.

Threshold wooden or metal element installed under a door and over the door sill and flooring; to prevent or block water and air from entering and to cover an open joint.

This chapter provides five routines which detail each element of tasks related to door installation and maintenance. Included are door hinge and lock installation, fitting a new door to an existing door jamb, and installing thresholds under doors.

SCOPE OF THE WORK

The work may range from 30 minutes to repair a hinge foundation to several hours for the installation of a new door. Only one person is needed for most jobs. The time-line plan for the job probably does not need to be developed because the project will, in all probability, be accomplished in one time period.

Because the tasks require so many subtasks and such a variety of tools, a detailed study of the task should be made. In addition, practice planing, chiseling out a mortise for a hinge, and drilling for a lock should be attempted until some degree of skill is achieved.

Above all, remember that the work is classified as finish; therefore, any errors will show.

In order to more fully appreciate the complexities of door maintenance and installation, this section provides a brief description of doors and their manufacture as well as numerous problem-solving tips, should problems arise.

DOORS AND THEIR MAKEUP

Figure 15-1 illustrates three distinct methods of door construction. Figure 15-1A shows a conventional *panel door*. Notice that it consists of styles and rails. The styles run full length and provide more than adequate foundation for hinge and lock installation. Figure 15-1B shows a *solid flush door*. Its interior structure consists of a series of strips of wood butted and glued to outer panels of veneered plywood. This type of door is used where exterior door needs apply. Figure 15-1C illustrates a *hollow-core door*. A frame of wood made from 1-1/2-in.-wide strips is used for the style and 2-1/2-in. strips for top and bottom; added blocks are included for lock installation. The remainder of the core may be either

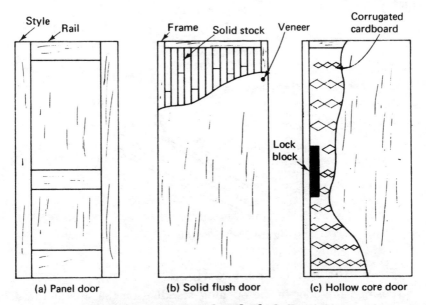

(a) Panel door   (b) Solid flush door   (c) Hollow core door

Figure 15-1 Panel and Flush Doors

empty or contain a series of corrugated cardboard reinforcements. To this frame thin sheets of veneered plywood are glued. These doors are classified according to their glue as interior or exterior.

FITTING AND HANGING A DOOR

Routines BDM1, Fitting a Door, and BDM2, Setting a Hinge and Hanging a Door, provide a detailed sequence of steps. Generally the sequence is very standard. A preliminary fitting must be made where the door will fit into the jamb. Following this the hinges are installed on the door, followed by installing them on the jamb. Once the door is swinging, final fitting is accomplished. To help you understand some of the finer points, let's examine each phase more closely.

Preliminary Cutting for Length

After laying the door across a pair of sawhorses to make cutting easier, measure the height of the jamb from floor to head. Mark and trim the bottom of the door so that, when cut, the door is 1/2 in. shorter than the jam height.

Veneer will splinter whether it is cut with a power saw or a handsaw. Therefore, lay a straight edge along the cutting line as shown in Figure 15-2 and score the veneer deeply with a utility knife or sharp wood chisel. When cutting with a power saw, score the *top side;* when cutting with a hand saw, score the *underside* of the door.

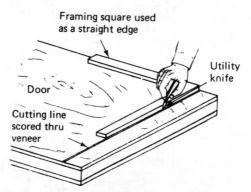

Figure 15-2 Scoring Veneer

Fitting the Hinge Side and Head

Position the door in the opening standing on the side of the door where the pins of the hinges will be visible when installed. Slip a flat bar

under the door and raise the door to the head and against the hinge side
of the jamb. Examine both the head and the side for fit. If needed, mark
areas to be cut or planed and take the steps necessary to make them fit.
Take the door down and install it in the door jack if one is made (Figure
15-3). If one is not made or used, prop the door *hinge side up* and prepare
the door back for hinges. Plane a bevel of 1/16 in. toward the inside (door-
stops side) the full length of the door. Mark the hinges for installation at
7 in. down from the top, 11 in. up from the bottom, and the center hinge
equidistant between the upper and lower hinges. Door hinges, called

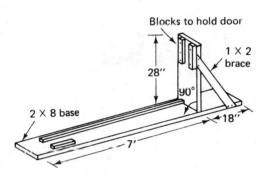

Figure 15-3 Door Jack

butts, are easily separated in two halves by removal of the pin. Proper
installation of the hinge may be seen by examining other doors or by
using Figure 15-4. Note in the figure how the hinge looks after installation. It fits snugly into a mortise area and is flush with the door's surface
edge. The mortise is cut by hand to a very close tolerance using very
sharp chisels. As the figure shows, the hinge does *not* extend to the inside
of the door surface except on 1-1/8-in. doors.

The top hinge on the jamb is set so that 1/8-in. clearance from door
top to jamb head results. This means, for example:

1. Distance from top of door to top of hinge   = 7 in.
2. Space needed for clearance   = 1/8 in.
3. Top of hinge on jamb from head down   = 7-1/8 in.

The hinge mortise (Figure 15-5) is marked and cut out and the
hinge half is installed with one screw. The door is hung on this hinge.
When the pin is installed, the other two hinges can be marked for upper
and lower position on the jamb. This technique of marking reduces the
chances for error.

When all hinges are installed and the door is swinging, the front

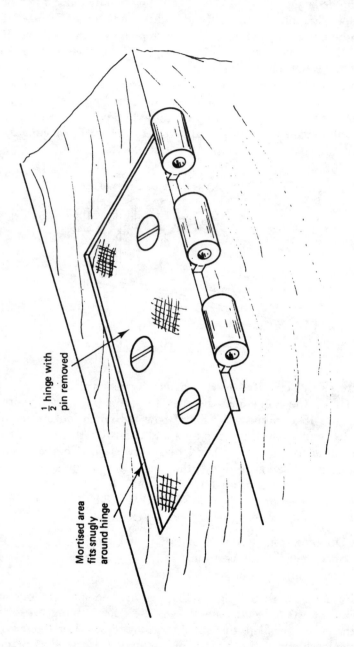

½ hinge with
pin removed

Mortised area
fits snugly
around hinge

Figure 15-4 Door Hinge

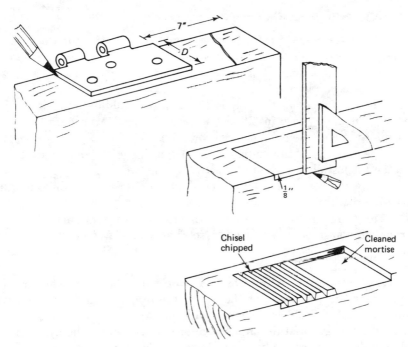

Figure 15-5 Setting a Hinge

of the door can be fit and adjustments can be made. Use Table 15-1 to locate your problem and the corresponding solution.

TABLE 15-1: What To Do if a Door Doesn't Fit

Problem	Solution
1. Space between door edge and jamb too wide	Mortise hinges at pin edge deeper.
2. Door top/head uneven or hitting	Deepen the lower and center hinge mortises or dress top of door.
3. Door binds on hinge side	Loosen screws on hinge and insert cardboard strip 1/4 in. x 3 in. x 1/16 in. thick behind hinge; tighten screws.
4. Door front doesn't close	Mark and dress down door edge; bevel toward stop.
5. Door looks good but leading edge hits jamb on closing and opening	Bevel the leading edge.

SETTING A LOCK

Instructions are provided with each lock set. If your door is pre-drilled (came installed in a door set), your task is simplified. On the other hand, if it requires boring for the parts of the lock, several factors need to be carefully studied.

The first recommendation is to practice installing a lock set in a 2 x 4 if you have never installed one. Practice making a mortise for the retaining plate of the cylinder and for the striking plate.

The second recommendation is to bore several holes with the expansion bit installed in the brace. Do this for two reasons: (1) to obtain the feel for the tool and obtain some degree of stability of direction, and (2) to verify the hole diameter needed.

The third recommendation is to lay out the retaining plate on a surface similar to the door's edge so that you can note while cutting the mortise the ease with which you can split, splinter, and otherwise destroy the surrounding wood.

Lock sets are usually installed 38 in. from the floor. The template is positioned to the inside of the door (the opposite side from the door stop) because this side is the long side of the bevel. The large hole or series of holes is bored through the door surface before the cylinder hole is bored. The proper method is to bore a hole until the tip of the bit or auger breaks through the opposite side. Remove the bit and finish boring the hole from the opposite side, using the breakthrough as a center guide. This technique prevents splintering of the veneer.

Bore the cylinder hole and follow by mortising the door edge for the retaining plate. Figure 15-6 shows an exploded view of the lock assembly and the carpentry required.

The striking plate, also shown in Figure 15-6, is needed to keep the door closed. Therefore, it is mortised into the jamb. When the plate is properly installed, the door will shut tight and not rattle. Two requirements must be met. The plate must be: (1) centered up and down to allow the plunger to enter the center hole, and (2) positioned in and out to eliminate door rattle. A combination square is the best tool to use for the layout part of the job.

Position the door so that it is almost closed. Mark on the jamb the upper and lower extremes of the plunger. With the square, draw two lines from the jamb edge to the door stop. Next, set the blade on the square by butting its end to the flat side of the plunger (Figure 15-7) and lock the blade. Scribe a line along the previous two marks made on the jamb while holding the square against the jamb.

Position the striking plate as shown in Figure 15-8 and mark its

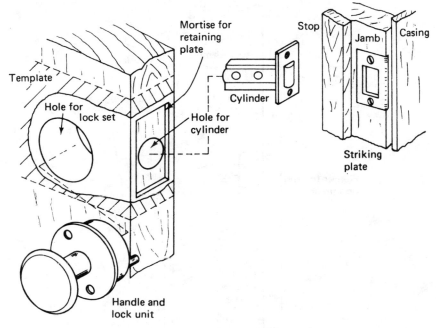

Figure 15-6 Door Prepared for Lockset

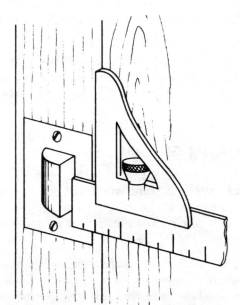

Figure 15-7 Finding the Depth of a Striking Plate

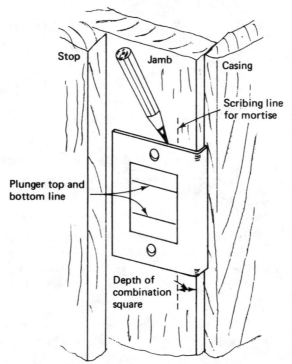

Figure 15-8 Marking the Outline of a Striking Plate

circumference. Mortise the entire area. Install the plate in the mortise and try closing the door. Adjust its position in or out as needed and install the screws. Use a 3/8-in. chisel and mortise the center-hole area to a depth of 3/8 in. or deeper if needed.

REPAIRING A HINGE AREA

Frequently the screws that hold a hinge in place work loose. If the holes are enlarged to a point where the screw cannot hold when tightened, specific steps must be taken to remedy the situation. Replacing the hinge with larger screws usually does not work well. Several alternatives are better (Figure 15-9):

1. Fill the hole with wooden match sticks. If the screw hole(s) is enlarged slightly, insertion of a match stick will fill the hole sufficiently to allow the screw to grab securely. This is a good technique on kitchen-cabinet doors.
2. Where the hole is greatly enlarged, a permanent method re-

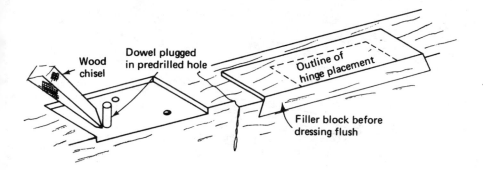

Figure 15-9 Repairing a Hinge Area

quires drilling the hole with a 1/4-in. or larger bit and plugging with glue and a dowel or plug. When the glue dries, the excess (plug) may be chiseled even with the mortise.

3. If the stock lumber around a screw hole has split, the split portion should be pried open slightly and glue forced into the crack. If possible, clamp the split area tightly and allow to dry. If it is impossible to use a clamp, drill a pilot hole and draw the split area tight with a screw.

4. If quite a large area of stock is rotted, cracked, or missing around a hinge, lay out and remove a fairly large segment of the stock. Custom-fit a filler piece, glue, and clamp it in place; after the glue dries, plane the surfaces and edges smooth to align with the original stock and mortise for the hinge.

FITTING A THRESHOLD

The last in this series of related tasks on door installation and maintenance is the fitting and installing of thresholds. Although there are many varieties of thresholds, you will probably encounter only two: those made of wood (oak) and aluminum (Figure 15-10). Their purpose is twofold: (1) they cover a joint between the door sill and the floor, and (2) they provide a form of insulation from the elements.

The wooden threshold (Figure 15-10B) is custom-fit to be centered under the door and extend past the jamb 3/4 in. on each end. The aluminum threshold is usually cut to fit snugly between the jamb. Routine BDM5 provides details for each type of threshold and its installation.

The wooden threshold should not be too difficult to install. It must

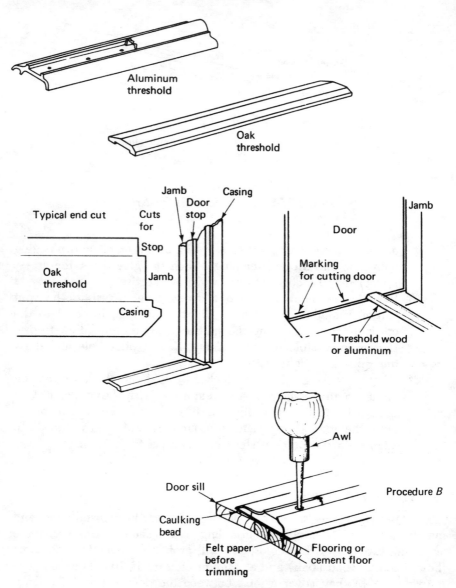

Figure 15-10 Thresholds

be carefully fit at each end, and predrilling for nailing or screwing is sometimes needed.

Aluminum thresholds (sometimes called *saddles*) present more work because they not only cover the joint between the sill and the floor but also insulate the bottom of the door from rain and wind. The rubber gasket must come into contact with the door bottom. Manufacturers' instructions usually specify the distance or space needed from the top of the metal to the door to allow for the gasket. They do not as a rule outline the waterproofing aspects.

A well-installed threshold will be set on at least one, and preferably two, strips of 30-lb felt paper. These strips should be cut 4 in. wide and trimmed after installation. Under the strip a bead of good-quality caulking should be laid. It should also extend along the door jamb for the width of the threshold. An additional bead of caulking should be laid on the under inside edge of the threshold. After the threshold has been screwed in place and the felt trimmed, glazing compound (white putty) should be forced into the corners at both ends of the threshold. Finally, the gasket should be installed.

Remember that a variety of aluminum thresholds are available. Study the manufacturer's instructions carefully. Add the extras that are provided here, and you should have a waterproof, windproof installation.

ROUTINES

Five routines are provided for door maintenance. Each details a part of a total task of replacing a door. However, each routine can be used separately.

BDM1: FITTING A DOOR

RESOURCES

Materials:
_____ door(s) _____ and _____
    no.   size   type

Tools:
1 no. 8 crosscut handsaw
1 block plane
1 no. 5-1/2 ripsaw
1 14- to 24-in. smoothing plane
1 6-ft folding ruler

1 straightedge
1 framing square

ESTIMATED MANHOURS

30 minutes to 1 hour

PROCEDURE

Step 1 Cut the excess ends of the styles from the door (if required).
a. Mark the cutting line with a framing square and pencil.
b. Cut off the excess with the saw.

Step 2 Measure the door-jamb height and determine if the door will fit in the opening. Mark and trim the bottom if required.

Step 3 Insert the door in the jamb and raise to the jamb head by inserting a flat bar under the door.

Step 4 Position the hinge side against the jamb and mark any areas that need planing. Mark the head for planing or trimming, as required.

Step 5 Remove the door from the jamb and position on the floor with the hinge edge *up,* against a corner, wall, or window, or insert in a door planing jack.

Step 6 Plane the *high* spots identified in step 4, then bevel the entire length 1/16 to 1/8 in. toward the *inside* of the door. (The inside is the closing direction.)

Step 7 Trim the head of the door according to the marks made in step 4.

Step 8 Reposition the door in the jamb.
a. Check the hinge side for fit (and plane as required).
b. Check the head for fit (and plane as required).

Step 9 Mark the front of the door for fit to the jamb.
Note: There should be 1/8 to 3/16 in. of clearance between the door edge and the jamb when the hinge edge is tight to the jamb.

Step 10 Cut and plane the door's leading edge. Bevel in the same direction as the hinge edge.
Note: Proceed with hinge installation.

BDM2: SETTING A HINGE AND HANGING A DOOR

RESOURCES

Materials:
2 hinges per door (1 pair) or 3 hinges per door (1-1/2 pairs)
Select one: 3-, 3-1/2-, or 4-in. butts

Tools:
1 set wood chisels
1 combination square
1 sharpening stone
1 2-in. phillips-head screw driver
1 automatic (yankee) screwdriver
1 8-in. common-point screw driver

ESTIMATED MANHOURS

30 minutes to 1 hour per door

PROCEDURE

Step 1 Lay out the hinge placement.
a. From the head, measure down 7 in.
b. From the bottom, measure up 11 in.
c. Center hinge between the top and bottom hinge.

Step 2 With the *pin* side of the hinge to the *outside* of the door, position the hinge against the line and mark the other end. Do this for all hinges.

Step 3 Position the hinge for depth equal to the edge of the main body of the hinge where the pin flange starts but *no closer to the opposite* door surface than 1/8 in. except on 1-1/8-in. doors. Mark the hinge width on the door. Adjust and lock the combination square for the hinge width marked in step 3 and mark all hinge areas.

Step   4   Reset the square for 1/8 in. or 1/8 in. full and mark the door face at the hinge area for depth of mortise.

Step   5   Cut out the mortise using a 1-1/4-in. chisel held erectly, with the flat side to the pencil lines. Cut the perimeters of the mortise.

Step   6   Chip the internal area of the mortise with successive cuts to the mortise depth.

Step   7   Clean out the mortise to the depth marked along the face of the door.

Step   8   Install the hinges in the mortises and secure each with *one* screw. Separate the halves of the top hinge by removing the pin.

Step   9   Measure from the head of the jamb on the hinge side of the jamb down 7-1/16 to 7-1/8 in. and draw a line.

Step 10   Position the hinge half against and below the line. Mark the bottom perimeter.

Step 11   Mark the depth of the mortise with a combination square and pencil. Reset for the width of the hinge's cutout to equal the door's cutout, and mark the jamb.

Step 12   Chisel out the stock to make a mortise.

Step 13   Install the hinge with one screw.

Step 14   Hang the door on the top hinge, hold the middle and bottom hinges against the jamb, and mark above and below each hinge. Remove the door, mark and mortise the jamb, and install the hinge.

Step 15   Hang the door; try for fit and proper closing as follows:

a. Check that the clearance between the top front and the back of the door to the jamb is 1/8 in.

b. Check that the door does not bind when in the closed position.

Adjust the hinge(s) accordingly.

Step 16   Install all screws in the door and jambs.

BDM3: SETTING A LOCK

RESOURCES

Materials:
_____ lock sets _____
    no.   type

Tools:

1 13-oz claw hammer
1 brace
1 expansion bit (to 3 in.)
1 6-ft folding ruler
1 set augers
1 set wood chisels
1 6-in phillips-head screwdriver
1 yankee screwdriver
1 6-in. common-point screwdriver
1 awl
1 combination square

ESTIMATED MANHOURS

45 minutes per lock set

PROCEDURE

Step 1 Measure and mark the leading edge of the door at a point 38 in. from the floor.

Step 2 Remove the template from the lock-set box and fold along the crease line; position the crease on the outside (high-bevel) edge of door and center on the 38-in. mark.

Step 3 Mark the center points for the cylinder and handle assembly with the awl or a 4d finishing nail.

Step 4 Set the adjustable expansion bit for the diameter of the hole needed through the door. Try one or more sample cuts in a scrap of 2 x 4 to certify that the diameter is exactly proper.

Step 5 Bore a hole through the door just until the *tip of the bit breaks through the opposite side.*

Step 6 Remove the bit from the door and complete the hole by boring from the opposite side, using the breakthrough point as a guide.

Step 7 Insert an auger of the proper size in the brace and bore the hole for the cylinder.

Step 8 Insert the cylinder in the hole, align the cylinder plate parallel to the door edges, mark the perimeter, and remove the cylinder.

Step 9 Using a 3/4-in. chisel, chisel a mortise around and within the perimeter just marked to a depth sufficient to allow the face plate to be *flush* with the door edge.

Step 10   Insert the cylinder with the plunger aimed properly (beveled surface pointing to the direction of door closing). Install two screws.

Step 11   Assemble the remainder of the lock according to instructions.

Step 12   Mark the jamb where the bolt strikes it.

Step 13   Set the combination square for depth by placing the body on the inside of the door and the blade against the flat side of the bolt. Lock the square.

Step 14   Mark a line on the jamb using the square from step 13. Position the leading edge of the cutout along the pencil line and centered (top to bottom) on where the bolt strikes the jamb. Mark the perimeter of the striking plate.

Step 15   Using a 1-in. chisel, make a shallow mortise to receive the striking plate. When fit, mark the center cutout area.

Step 16   Using a 1/2- or 3/8-in. chisel, mortise the stock in the center area of the striking plate along the lines drawn in step 15. Make a hole 3/8 to 1/2 in. deep.

Step 17   Use screws to install the striking plate. Try the door for proper closing.

BDM4: REPAIRING A HINGE AREA

RESOURCES

Materials:
Glue
Dowel(s)

Tools:
1 drill-and-bit set
1 1/2-in. wood chisel
1 13-oz claw hammer
1 utility knife
1 no. 8 crosscut handsaw
1 C or bar clamp

ESTIMATED MANHOURS

30 minutes per hinge

PROCEDURE

Step 1   To repair a hinge screw, remove the hinge to expose the problem area.

Step 2   Clean the surface or mortise and examine for defects.

Step 3   Insert the bit in the drill and drill a hole larger and deeper than the existing hole.

Caution: If drilling on a door, do *not* drill through the door.

Step 4   Apply glue in the hole(s) and drive a dowel into the hole. Allow to dry.

Step 5   Trim the excess dowel flush with the hinge mounting surface.

Step 6   For splits or cracks, clean the area to get rid of loose, rotten wood, dirt, and debris.

Step 7   Cut and tailor-fit a filler piece, if needed.

Step 8   Glue the split area and/or filler piece and split and clamp with a C or bar clamp.

Optional Step 8   Drill a pilot hole and countersink and install a screw to reinforce the split area. *Do not* install a screw where screws from the hinges will be driven.

BDM5: FITTING A THRESHOLD (WOODEN OR ALUMINUM)

RESOURCES

Materials:

_____ thresholds, oak _____
    no.   size

_____ thresholds, aluminum _____
    no.   size

Wood screws 1-1/2 in. × no. 10 or 1-3/4 in. × no. 10 FH
1 tube caulking
3 sq ft 30-lb felt paper
1/2 pint glazing compound
6 8d galvanized finishing nails

Tools:
1 13-oz claw hammer
1 1/16-in. nail set
1 hacksaw
1 yankee screwdriver
1 10-in. common screwdriver
1 10-in. cross-point screwdriver
1 3/4-in. wood chisel
1 power saw
1 awl
1 caulking gun
1 utility knife
1 block plane
1 6-ft folding ruler
1 no. 8 crosscut handsaw
1 drill-and-bit set

ESTIMATED MANHOURS

1 hour

PROCEDURE A: WOODEN THRESHOLD

Step 1 Measure and cut the threshold 1-1/2 in. longer than the distance from jamb to jamb.

Step 2 Position the threshold so that the flat surface is at a point centered under the door. (*Note:* The door must be fully open or off its hinges.)

Step 3 Use a pencil to mark the threshold on either side of the jamb's edge and the door trim.

Step 4 Lay the threshold flat and center against the door trim and the jamb. Mark a cross-cut for the jamb cutout and door-trim cutout. Connect all the marks with pencil lines and the combination square.

Step 5 Repeat steps 3 and 4 for the other end of the threshold.

Step 6 Close and/or rehang and close the door.

Step 7 Lay the threshold flat on the floor with the end against the door. With awl or pencil, scribe lines along the flat surface of the threshold on the door near both front and rear edges and at the door's center.

Step 8 Cut away the stock on the threshold according to the marks previously made.

Step 9 Remove the door from the jamb and install the threshold.

Step 10 If no adjustments need to be made, drill several pilot holes through the threshold (four holes) and countersink for the screwheads.

Step 11 Screw the threshold to the door sill.

Step 12 Connect the lines on the door. Cut off the bottom and dress the edges with the block plane.

PROCEDURE B: ALUMINUM THRESHOLD

Step 1 Measure and cut the threshold exactly long enough to fit between the jambs.

Step 2 Butt the end of the threshold to the door (close the door) and mark the door along the flat side of the threshold at the ends and center of the door.

Step 3 Remove the door to a workbench or saw horses.

Step 4 Measure up 1/8 in. from all marks made in step 2 and connect lines. Cut off the door bottom and smooth the edges with the block plane.

Step 5 Cut one or two strips of felt paper 4 in. wide × the length of the threshold.

Step 6 Lay felt strips over the door sill and floor, then the threshold over the felt.

Note: If the threshold has a rubber bumper, remove the bumper and ensure clearance of 1/8 in. between the door bottom and the threshold. If the threshold has no bumper, make it fit as close as possible, but *not* striking.

Step 7 Adjust the felt paper (add some or take some away) until the threshold is aligned properly.
For concrete floors, skip to step 11.

Step 8 Drive two 4d finishing nails along the outer edge of the threshold. Remove the threshold but not the felt.

Step 9   Caulk the ends and *inner* edge line of the threshold. Embed the threshold in caulking.

Step 10   Screw the threshold to the door sill (if wooden).

Step 11   For concrete floors, open the door fully and drive a 4d finishing nail into the jambs at the outer edge of the threshold (marked).

Step 12   With an awl, mark the concrete for the drilling of holes for plugs into which to screw the threshold.

Step 13   Remove the threshold and felt and drill holes 3/8 or 1/2 in. x 2 in. deep. Fill with wood plugs trimmed flush with cement.

Step 14   Reposition the felt and caulk along the insides of the jamb and threshold line. Embed threshold in caulking and align against the 4d finishing nails.

Step 15   Screw the threshold to the floor, trim the felt with the utility knife, putty the ends of the threshold, and, if required, insert the rubber bumper.

16

SIDING

BASIC TERMS

Battens 1 x 2, 1 x 3, or 1 x 4 stock lumber nailed on center at right angles to studs, or vertically over joints in siding.

"Breather"-type building paper building paper designed to block moisture but allow limited passage of air.

Course horizontal row of siding, shingle, or outer covering on a wall.

Forming method of molding window and door-cap flashing to rain-cap molding.

S4S sanded or smoothed four sides.

Shim tailormade wedge that fits behind a lapsiding member, usually at a joint, to provide reinforcement.

Starter strip narrow piece of lumber, usually installed at the foundation line, used to tilt the first course of siding to the proper bevel.

Story pole layout piece of 1 x 2 or 1 x 3 which guides installation of courses of siding.

When you have reached this point, you are well on the way to performing *finishing* work. This type of work is time-consuming and exacting. The quality of the job lies not only in the quality of the material but in the quality of the work performance.

Tools are primary objects in doing a good job. This chapter discusses many of these tools. There are twelve routines to cover the various types of siding and to describe subtasks that are used but not seen in almost every siding job. Overviews of siding jobs are presented, showing the uses of the subroutines with the various types of siding.

SCOPE OF THE WORK

Certain types of siding, such as shingles and narrow lapsiding, may be installed by one person. A helper is usually desirable, however. For wide lapsiding, vertical siding, and exterior paneling, a helper is essential to hold and nail and frequently to mark boards.

After reviewing the routines at the end of the chapter you may be surprised to learn how few square feet can be installed per hour. Don't be discouraged; there are simply many phases involved in proper installation. Each piece must be measured; the ends trimmed; marked for cutouts, bevels, etc.; drilled and cut as required; and nailed. Added to these tasks are those of preparation, such as waterproofing the walls and making and/or installing corners and flashing.

Also included but not specified is the continuing requirement for scaffold erection, whether by sawhorses or ladders or by 2 x 4s and 1 x 6 ledgers.

The selection of the type of siding will in part dictate the processes used. However, before any type of siding can be installed, a few preparatory tasks need to be performed.

SEALING THE WALLS

Some walls may have been sheathed with tar-impregnated materials (fiberboards). If so, installation of building paper may be limited to covering the joints of the fiberboard. As a rule, though, either tar-impregnated felt (15 or 30 lb) or "breather"-type building paper is first installed over the entire outer surface of the wall. The simplest method of installation is to precut the roll into 8- to 12-ft strips and allow them to lie flat. Nail each strip on the wall while keeping it straight, without bulges and wrinkles. Nail it with 1-in. roofing nails. Overlap the ends by 4 in. and the parallel edges by 2 to 4 in. Be sure to bring the paper around the corners a minimum of 8 to 12 in. With the building paper installed and trimmed closely around doors and windows, you are ready for the next preliminary step.

INSTALLING FLASHING OVER DOORS AND WINDOWS

Metal flashing must be installed over each window and door. Recall from Chapters 13 through 15 that most units are purchased with drip caps installed. If your unit does not have a drip cap, it is recommended that you buy and install one. Figure 16-1 shows the flashing installed on a drip cap over either a window or door. Metal, probably aluminum, must be purchased from your local lumberyard to do the job. It is sold by the running foot with widths starting at 12 in. Unless you have a specific reason to do otherwise, a 12-in.-wide strip should be split in two because this will give you enough metal for two windows or doors. Measure the drip-cap length and *add* 1-3/4 in. when cutting the metal for length. The

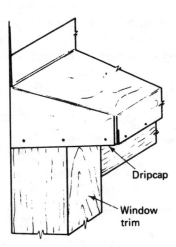

Figure 16-1 Flashing Cap Molding on Windows and Doors

added length is needed to seal the ends of the drip cap and the wall area to prevent the entry of blowing rain.

Hold the machine edge even with the bottom of the drip cap. Drive a corrosion-resistant nail (aluminum for aluminum flashing) about midway from the ends into the drip cap (see Figure 16-2). Space the nails evenly and at 1-1/2 to 2-1/2 in. apart; work first on one side of the center, then on the other side, until you reach the ends.

Follow the nailing with forming by using a block of 2 x 4 12 in. long and a hammer to fold the metal to the slope of the cap and the wall (Figure 16-2). When the flashing is molded, again working from the center out, press the molding to the cap and wall and nail it near the top into the wall. Four or five nails will hold it.

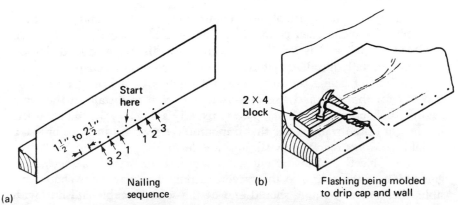

Figure 16-2 Forming Flashing to Cap

Look closely at Figure 16-3; the three steps show how to cut and form the ends of the flashing. Snip the areas indicated. Fold the front piece back. Press the top piece over it and nail. With all the window and door caps flashed, and the building paper installed, the general tasks are complete. The layout of siding must be considered next.

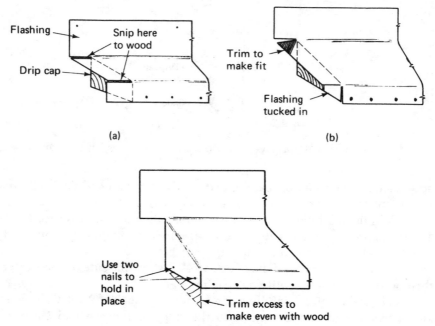

Figure 16-3 Folding Flashing over Ends

LAYOUT OF THE STORY POLE

The story pole is a tool most commonly used by a mason. With it he is able to define each course of block or brick. Because it has been used for so long and is a fail-safe method for alignment and definition of all horizontally installed material, it is discussed in this chapter.

Instead of laying the story pole out for blocks or bricks, it is to be laid out for the type and size of siding you plan to install on the wall. Figure 16-4 shows that a story pole is usually a 1 x 2 or 1 x 3 cut to the full height of the wall. Notice that our sample shows graduations of 9-1/2 in. and markings for window sill and header.

The objective of laying out a story pole is to provide a specific organization of the siding so that several things are planned. Where possible, the course of siding should end at the soffit or gable-end line with

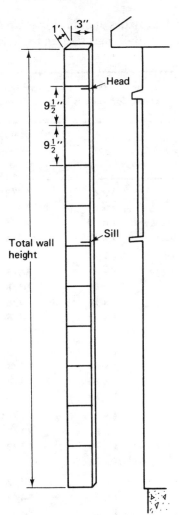

Figure 16-4 Story Pole

close to a full width. Also, it is desirable to have a course start *even* with the window sill to preclude cutting narrow strips of siding below the window sill. Another detail of the plan is to have the last course alongside a window and its top edge extend the minimum overlap distance *above* the drip cap. This will provide a basis for the installation of an uncut straight siding board or an uncut straight frieze board over the window.

Each type of siding has a minimum recommended overlap. However, the overlap may be exceeded by one-half the recommended quantity with no adverse effects. These are some of the reasons for laying out a story pole. Table 16-1 provides standard overlap for various types of siding, (FACE) and Table 16-2, nail size and spacing.

TABLE 16-1: Coverage Estimator   This estimator provides factors for determining the exact amount of material needed for various types of siding.

		Width (in.)		
	Nominal Size (in.)	Dress	Face	Area Factor
Shiplap	1 x 6	5-1/2	5-1/8	1.17
	1 x 8	7-1/4	6-7/8	1.16
	1 x 10	9-1/4	8-7/8	1.13
	1 x 12	11-1/4	10-7/8	1.10
Tongue and groove	1 x 4	3-3/8	3-1/8	1.28
	1 x 6	5-3/8	5-1/8	1.17
	1 x 8	7-1/8	6-7/8	1.16
	1 x 10	9-1/8	8-7/8	1.13
	1 x 12	11-1/8	10-7/8	1.10
S4S	1 x 4	3-1/2	3-1/2	1.14
	1 x 6	5-1/2	5-1/2	1.09
	1 x 8	7-1/4	7-1/4	1.10
	1 x 10	9-1/4	9-1/4	1.08
	1 x 12	11-1/4	11-1/4	1.07
Paneling patterns	1 x 6	5-7/16	5-1/16	1.19
	1 x 8	7-1/8	6-3/4	1.19
	1 x 10	9-1/8	8-3/4	1.14
	1 x 12	11-1/8	10-3/4	1.12
Bevel siding (1-in. lap)	1 x 4	3-1/2	2-1/2	1.60
	1 x 6	5-1/2	4-1/2	1.33
	1 x 8	7-1/4	6-1/4	1.28
	1 x 10	9-1/4	8-1/4	1.21
	1 x 12	11-1/4	10-1/4	1.17
Asbestos siding	12 x 24	24	11	1.09
		24	10-1/2	1.14
Wood shingles	3/8 x 18	Random	5	3.6
			6	3.0
			8	2.5

To compute the square footage, select a type and size of siding. Measure the height and length of the wall(s) and multiply H × L × factor. Add an allowance for trim and waste (approximately 10 percent). The length of a shingle does not alter the area factor figure.

TABLE 16-2

Framing	Nail Size and Spacing	
Maximum 24 in. OC	When racking requirements of FHA Circular 12 must be met:	When racking requirements of FHA Circular 12 need not be met:
	6d or 8d galvanized box nail 3/8 in. in from panel edges 4 in. OC along all edges 8 in. OC along intermediate supports	6d or 8d galvanized box nail 3/8 in. in from panel edges 6 in. OC along all edges 12 in. OC along intermediate supports

Use 6d box nails only for direct panel-to-stud applications or for panel-over-wood or plywood sheathing applications.

CORNER PREPARATION

Corner boards and metal corners are frequently installed to make special effects and to lessen the work. Inside corners relieve the need for custom-fitting pieces of lapsiding. Outside corners relieve the need for custom-fitting lapsiding. A wooden outside corner is made from two pieces of stock: one 1 x 4 and one 1 x 3. When nailed together as shown in Figure 16-5A, they form a corner whose sides are 3-1/2 in.

Some sidings, asbestos, for instance, generally require a metal outside corner to make a waterproof corner. They may use a metal inside corner as well (Figure 16-5C and D). This type of corner is carefully nailed on each corner after building paper is installed.

HORIZONTALLY INSTALLED SIDING

Install necessary inside corners (1-1/8 inches thick) for the full height of the wall to 1/8 in. below the sheathing. Install outside corners, if required. Nail a level starter strip of 3/8 in. x 1-1/2. in. wood along the bottom edge of sheathing (Figure 16-6).

Measure on a wall the width of the siding at several points from the bottom of the sheathing, and snap a chalk line. Cut and prepare siding for installation. Hold the top edge of the siding even with the chalk line, and nail at least 1/2 in. from the bottom edge through the starter strip, at the siding ends, and at each stud location.

Butt joints should occur only at stud locations and should be nailed

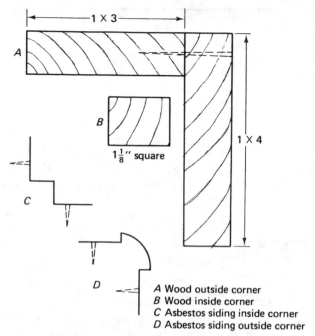

A Wood outside corner
B Wood inside corner
C Asbestos siding inside corner
D Asbestos siding outside corner

Figure 16-5 Inside and Outside Corners

at both top and bottom. Adjacent pieces should touch lightly at butt joints and must never be forced or sprung into place.

Second and subsequent courses of siding should be applied after their positions are established by use of the story pole and chalk line. All splices should be made in different locations from the course below.

Each piece of siding must be individually tailored around windows and doors. Fits should be close but not forced. Some products require 1/8-in. spacing, so follow the manufacturer's directions, if provided.

Shims are sometimes needed. If S4S type 1 x 10 or 1 x 12 siding is used, a wedge-shaped shim is essential to each splice/joint to keep both pieces flush. Shims may also be needed above the drip-cap molding to provide a solid base.

The self-returning corner may be used on a variety of siding, including cedar shingles. It is made by beveling first one piece of siding, then beveling the adjacent piece to fit (Figure 16-7). The top edge of siding is flush with the wall. The bottom edge is flush with the outside surface of the previously installed course.

The corner is completed by installing the siding from the adjacent wall (same course) overlapping the corner. After nailing in place, or

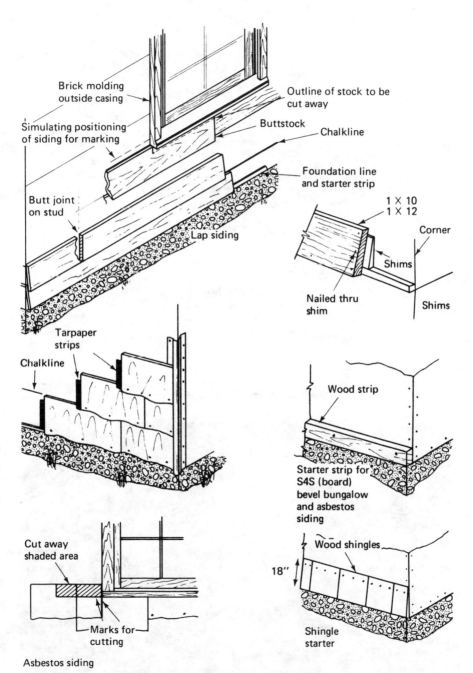

Brick molding outside casing

Outline of stock to be cut away

Simulating positioning of siding for marking

Buttstock

Chalkline

Foundation line and starter strip

Butt joint on stud

Lap siding

1 × 10
1 × 12

Corner

Shims

Nailed thru shim

Shims

Tarpaper strips

Chalkline

Wood strip

Starter strip for S4S (board) bevel bungalow and asbestos siding

Cut away shaded area

Wood shingles

18″

Marks for cutting

Shingle starter

Asbestos siding

Figure 16-6 Siding Installation Tips

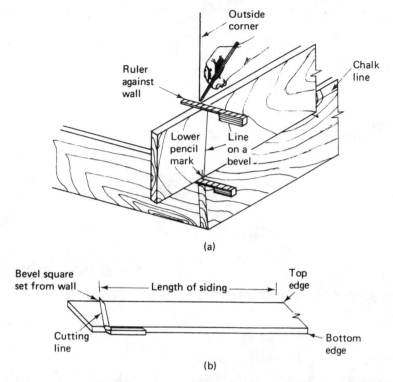

(a)

(b)

Figure 16-7 Self-Return on Corners

Figure 16-8 Using Metal Corners   *(Photo courtesy of Masonite Corporation)*

marking, cutting, and nailing, the corner is trimmed smooth with a block plane.

Metal corners may be used as self-return corners when desired (Figure 16-8). Once again, see the manufacturer's instructions.

VERTICALLY INSTALLED SIDING

Many types of siding may be installed vertically: S4S in various widths, channel ship-lap, and various tongue-and-groove designs with beveled or molded edges.

Preparatory work must include providing nailing surfaces either installed as cats (horizontal blockings) within the 2 x 4 studs at 16-in.-OC intervals from floor to ceiling, or 1 x 4 battens nailed over the fiberboard sheathing. Generally, when vertical siding is to be installed, solid wood sheathing is not used. Fiberboard is used.

The main problem area to be aware of and continuously verify is that of maintaining a vertical line that is *plumb*. Before installation begins and after the building paper is installed, vertical chalk lines should be snapped at 4-ft intervals. During the installation these chalk lines provide a continuing reference point from which to maintain a plumb line (Figure 16-9).

The method of nailing, especially tongue-and-groove types, will also contribute to alignment. Alternately, nail one piece from the top to the bottom, one from the bottom to the top, and one from the center to the top and bottom. This method prevents bulging and compacting at bottom, center, or top.

Where S4S plank types are installed (Figure 16-10), the compacting problem does not exist. However, slight bows and warps in the stock must be removed by the nailing processes. The bow may be easily defined by sighting the board prior to installing. If the bow is near an end, start nailing at that end. If the bow is in the middle, start nailing at the midpoint and work toward the ends.

Cutouts for windows and doors must be individually tailored. Seldom, if ever, is an outside corner or inside corner required since the siding is being installed vertically. No starter strips need to be used.

VERTICALLY INSTALLED PANELS

The final type of siding discussed in this chapter is exterior panel installation. Panels, usually sold in 4 ft x 8 ft and 4 ft x 9 ft sheets, are available from a number of manufacturers. Panels are fairly easy to install but require a fair degree of accuracy. Errors resulting in a wasted sheet

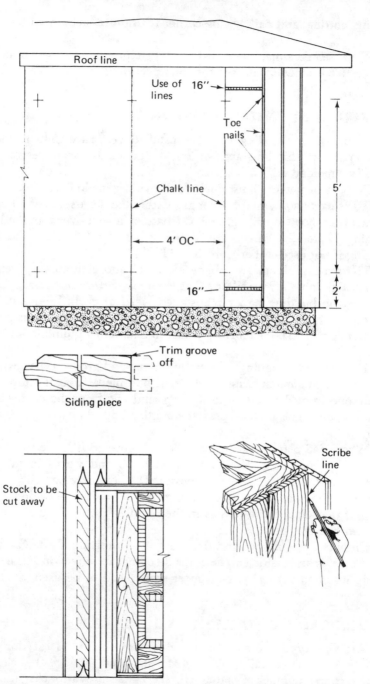

Figure 16-9   Cutting, Nailing, and Installing Tongue-and-Groove Siding

Figure 16-10 Vertical-Board Siding *(Photo courtesy of Masonite Corporation)*

Figure 16-11 Panel Siding *(Photo courtesy of Masonite Corporation)*

may cost $10.00 or more. Preliminary steps must be taken prior to installing the panels: sealing the walls with vapor-barrier paper as required by local ordinances and laying out how the panels will be installed.

Install panels so that all edges and joints fall on studs and so that no piece of 6 in. or less in width is used. All butted joints should touch gently but should not be forced. All cuts around windows and doors should be made to allow approximately 1/8 in. spacing, which can be caulked later.

If the panels are not made with a tongue and groove, they will probably be battened. They may also be sealed without battens. To join two panels without the batten, cut the edge on joining sheets to 45 degrees (one open and one closed). Install the open end cut sheet first, caulking the joint and the stud area behind the joint. Press the closed cut edge into the caulking, align it properly, and nail (Figure 16-12).

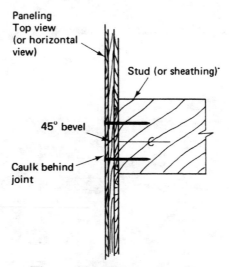

Figure 16-12 Joining Panels

Layouts for window and door openings should be made on the back-side of the panel if a power saw is to be used for cutting. If a hand saw is to be used, the layout should be made on the face side of the panel. The layout is generally established from two reference points. One is the sill line or the lookout ledger, and the other is the last sheet installed or corner reference.

Using a ruler, measure the distance from the various references to the window or door or corner (final sheet). Position the sheet to be cut so as to approximate its final position. Transfer the measurements and join the lines with the aid of a straightedge. Predrill several holes at the corners *inside the waste area* with a brace and a 3/4- or 1-in. auger. Start the saw cut with a keyhole saw and finish with a fine-toothed saw (no. 8 or finer). Install the sheet in place, nail, and caulk around the casing.

ROUTINES

The first few routines of the twelve that follow outline preparatory tasks associated with a variety of siding applications. Following these are routines that outline specific types of siding installation techniques for such sidings as asbestos, lap, vertical, tongue-and-groove, and panels.

BSid1: LAYING OUT SIDING

RESOURCES

Materials:
1 1 x 2 or 1 x 4 10 ft long
Elevation and/or detail drawings from plan
Type of siding _____ quantity _____
sq ft
Size of siding _____ (see Tables 16-1 and 16-2)

Tools:
1 6-ft folding ruler
1 stepladder or extension ladder (optional)

ESTIMATED MANHOURS

1 hour

PROCEDURE

Step 1 Measure the height of the wall from the top of the foundation to *either* 1 to 1-1/2 in. above the window casing or to the lookout ledger.

Step 2 Transcribe the length from step 1 to the story pole and cut off at the mark.

Step 3 Measure the height of the window sill's lower surface from the foundation and mark on the story pole.

Step 4 Measure the window/door top of casing/rain drip from foundation and mark on the story pole.

Step 5 Use Table 16-1 (face width) for recommended exposure of siding.

Note: Lay out the story pole on a wall with a window. If there is no window, pick a wall with a door. If there is no window or door, use a blank wall.

Step 6 Using the recommended exposure and priority cited above, space off, from the bottom of the story pole, courses of siding. Adjust the spacing between the minimum/maximum allowances until the rows space out evenly, plus one additional overlap above the window drip cap/casing.

Note: Try to space the area from the foundation to the sill area so that no cutting of siding is needed under the sill. Slight variations in exposed distances will not be seen.

Step 7 Mark the rows with crayon or notch with a saw. (If chalk lines are desired on the wall area instead of on the *last course installed,* use step 8.)

Step 8 Make one more set of marks on the story pole equal to the overlap width above the marks cut in step 7. Use this series of marks to mark the wall for chalk lines. The siding will be placed with its top edge even with the chalk line.

Step 9 Make a duplicate story pole so that each person in a two-person installation will have a story pole.

BSid2: INSTALLING BUILDING PAPER

RESOURCES

Materials:
One roll of building paper for each 400 sq ft of wall surface (use the same square footage of wall surface as used for wall sheathing)

3 lb 1-in. galvanized roofing nails per roll of felt

Tools:
1 13-oz claw hammer
1 utility knife
1 sawhorse or ladder (optional)

ESTIMATED MANHOURS

For two people   2 hours per roll

PROCEDURE

Step 1   Precut strips of paper into 8- to 12-ft strips; lay them flat.

Step 2   Nail strips onto the wall, overlapping each end by 2 to 4 in. Place the second rows over the first, overlapping 4 in.

Step 3   Nail the felt at 6-in. intervals along the edges and at 12-in. intervals through the center.

BSid3: INSTALLING CORNER BOARDS

RESOURCES

Materials:
_____ 1 x 4 × 8 or 10 ft no. 2 common or better grade (circle one)
   no.

for each corner
_____ 1 x 3 × 8 or 10 ft (circle one) for each corner.
   no.

1/2 lb 8d galvanized finishing nails per corner _____
                                       total needed

Tools:
1 13-oz claw hammer
1 no. 8 crosscut handsaw

1 pair sawhorses
1 6-ft folding ruler

ESTIMATED MANHOURS

20 minutes per corner

PROCEDURE

Step 1   Measure the total height of the wall from the foundation to the lookout ledger or the lower planed edge of the frieze board.

Step 2   Cut a 1 x 4 and a 1 x 3 for each corner.

Step 3   Nail a 1 x 3 at 12-in. intervals to the corner with its outer edge even with the corner.

Step 4   Nail a 1 x 4 to the 1 x 3 and stud, completing the corner board. *Note:* Reverse steps 4 and 3 for an inside corner, or use a 1-1/8-in.-square stock.

Step 5   Set all nails with a 1/8-in. nail set.

BSid4: INSTALLING STARTER STRIPS

RESOURCES

Materials:
_____ lineal ft starter-strip material (type of material _____)
    no.

Size 1/4 in. 3/4 in. × 1-1/2 in. 2 in. (select the size needed)
_____ lb nails. One pound is sufficient for 30 lineal ft of starter strip.

Tools:
1 13-oz claw hammer
1 no. 8 crosscut handsaw
1 combination square
1 power saw (optional)
special-purpose cutter for aluminum and asbestos (optional)

ESTIMATED MANHOURS

15 minutes per 12 ft

PROCEDURE

Step 1 Precut starter strips to the size selected in the materials section.

Step 2 Nail the starter strips even with the foundation line on the wall.

BSid5: INSTALLING FLASHING OVER DOORS AND WINDOWS

RESOURCES

Materials:

1/4 lb 3d galvanized common nails per window/door _____ ✕
                                                      no.

1/4 lb = _____ lb.
           total

Flashing metal aluminum/galvanized: 12-in. roll ✕ _____ feet. Each
                                                    length

piece split in half produces two pieces.

Plans

Tools:

1 pair snips
1 13-oz claw hammer
1 block of 2 x 4 12 in. long
1 straight edge

ESTIMATED MANHOURS

45 minutes per window/door

PROCEDURE

Step 1   Measure the length of the window/door drip cap or casing.

Step 2   Cut a piece of flashing equal to the length in step 1 *plus* 1-3/4 inches.

Step 3   Mark the center line on the flashing using the straightedge as a guide. Cut the flashing in two pieces (lengthwise).

Step 4   Center the flashing along the drip cap. Starting from the center, nail flashing at 1-1/2- to 2-1/2-in. intervals.

Step 5   Using a 2 x 4 block and hammer, mold the flashing over the drip cap and up on the wall.

Step 6   Hold the flashing down on the drip cap and nail to the wall at the top lip of the flashing. Four nails will usually do the job.

Step 7   Clip bends on the end of the flashing and fold the overhang in and down. Nail.

Note: On casings without the drip cap, either install a drip cap or extend the flashing 1/2 in. down from the top edge of the casing and nail as in step 4.

BSid6: INSTALLING LAPSIDING

RESOURCES

Materials:
Lapsiding requirements (see Table 16-1)

    Type _____

    Size _____

    Square feet _____

Nails

    Type _____

    Size _____

    Quantity _____ lb. (at 1 lb per 50 lineal ft)

Tools:
1 framing square

1 bevel square
1 combination square
1 no. 8 crosscut handsaw
1 13-oz hammer
1 chalk line
1 pair sawhorses
1 1/8-in. nail set
1 story pole (optional)
1 scaffold (optional)

ESTIMATED MANHOURS

For two people   5 sq ft of wall surface per hour

PROCEDURE

Preliminary steps   Install building paper, starter strip, corner boards if needed, flashing, and make story pole.

Step   1   Trim the end of a piece of siding by squaring and cutting it.

Step   2   Measure up from the foundation and mark the width of the siding board, less the amount of siding to extend below the sill/foundation line.

Step   3   Snap a line along the marks made in step 2.

Step   4   Position the first board and mark the other end for cutting or nail in place.

If self-corners are used, study Routine BSid 7 before proceeding with step 5.

Step   5   Install the first course on the adjacent wall, making the corner as in Routine BSid7.

Step   6   Use the story pole to mark the position of the bottom (or top) edge of the second course. Snap a chalk line.

Step   7   Precut one end of another board and position it on the wall.

Note: If a splice is needed, *break* the joint from the first row. Place a wedge-shaped shim behind the joint before nailing. The break must be on a stud.

Step   8   Measure and cut the other end. Nail in place.

Step   9   For window or door cutouts, hold the board under (or over) the window to mark it for cutting.

Step 10 Measure from the story-pole mark made on the last course of siding installed to the window sill or head or the door's head.

Step 11 Lay out the measurement from steps 9 and 10 on the siding to be installed. Cut the stock away. Install the board.

BSid7: MAKING A SELF-CORNER ON LAPSIDING

RESOURCES

Materials:
None required

Tools:
1 6-in. block plane
1 bevel square
1 no. 8 or 10 crosscut handsaw
1 13-oz claw hammer
1 1/8-in. nail set
1 wooden folding ruler
1 pair sawhorses (optional)

ESTIMATED MANHOURS

15 minutes per corner

PROCEDURE A: TRIMMING CORNERS ON THE WALL

Preliminary step Install a course of siding to allow its end to extend past the corner.

Step 1 Using your ruler, make a mark on the outside surface (top and bottom) of the siding even with the wall's corner (top), and either the starter strip or the last course installed (bottom).

Step 2 Draw a line connecting the two marks made in step 1. *The line must be on a bevel.*

Step 3   Cut the excess stock away so that the pencil line remains on the nailed portion of the siding.

Step 4   Allow siding from the adjacent wall (same course) to extend past the corner. Nail in place.

Step 5   Mark the siding top and bottom as in steps 1 and 2.

Step 6   Cut off the excess stock.

Step 7   Plane cut the surface smooth with the block plane.

PROCEDURE B: PRECUTTING BEVELS ON SAWHORSES

Step 1   Hold a board in place on the wall and mark the lower edge where it passes either the starter strip or the last course installed.

Step 2   Set the bevel square for the bevel needed on the cut.
a. Set the square's handle on the starter board or first course.
b. Have one end of the blade flush with the starter board or first course and the other end of the blade flush with the wall corner. Tighten the blade.

Step 3   Use the bevel square, set, and mark the siding measured in step 1.

Step 4   Undercut the marked siding.

Step 5   Install on the wall and nail in place.

Step 6   Repeat steps 1 and 3 through 5 for the adjacent piece of siding.

Step 7   Plane-cut the edge smooth with a block plane.

BSid8: CUTTING, NAILING, AND INSTALLING ASBESTOS SIDING

RESOURCES

Materials:
Asbestos siding requirements (see BSid 1 if used):
_____ of bundles. One bundle covers 33 square feet.
   no.

2 lb nails per square

Outside corner metal _____ of 8 ft
no. pieces

Inside corner metal _____ of 8 ft
no. pieces

Tools:
1 asbestos cutter (rented)
1 13-oz hammer
1 chalk line
1 pair sawhorses
1 pair snips
1 hacksaw

ESTIMATED MANHOURS

6 sq ft of siding per hour

PROCEDURE

Preliminary steps   Install breather-type sheathing paper on the wall surface, lay out the story pole, and install flashing over the doors and windows.

Step 1   Cut and install inside and outside corner strips. Nail strips accurately and evenly over the corners.

Step 2   Precut 1/4 in. × 1-1/2-in. stock pieces and nail as starter strips.

Step 3   Measure up from the foundation/sill line 12 in. less the amount of shingle overlap of foundation line desired (usually 1/2 in.).

Step 4   Snap a chalk line along the marks made in step 3.

Step 5   Starting from a corner, install full pieces of siding; nail through predrilled holes in the bottom of the shingle. Insert a tarpaper strip under each vertical joint.

Step 6   Hold the shingle to be cut in place and mark. Insert in the cutter and cut off. Predrill nail holes as required. Nail in place.

Step 7   Use the story pole to mark off the second and higher rows. Snap the chalk lines.

Step 8   To cut around doors and windows, mark the shingle for the area to be cut away. Use the chipper cutter on the cutting machine to chip along the lines drawn. Prepunch new holes. Nail in place.

Step 9 Caulk the joint along the window and door casings. (Use Routine BSid12, if necessary.)

BSid9: CUTTING, NAILING, AND INSTALLING SHINGLES

RESOURCES

Materials:
Underlayment shingles (unless calculated in Routine BSid1)
No. squares _____ of 3/8 in. × 18 in. cedar
No. squares _____ exterior layer: size _____, type _____
_____ lb 4d common galvanized/corrosion-resistant nails
    no.
_____ lb 6d box galvanized/corrosion-resistant nails
    no.

Tools:
1 no. 8 crosscut handsaw
1 13-oz hammer
1 chalk line
1 story pole
1 framing square
1 pair sawhorses
1 hand ax and/or block plane
1 scaffold or ladder (optional)

ESTIMATED MANHOURS

2 sq ft per hour

PROCEDURE

Preliminary steps Install building paper on the side wall (Routine BSid2). Install flashing over the doors and windows (Routine BSid5). Lay out the story pole (Routine BSid1).

Step 1   Nail a course of underlayment shingles across the wall with the bottom edges even with the sill/foundation line.

Step 2   Install the top layer over the underlayment course, breaking the joint. Nail above the exposure line.

Note: Use Routine BSid7 as a guide for making corners.

Step 3   Use a story pole and mark the second course on the shingle surface and snap a chalk line.

Step 4   Nail a course of underlayment shingles 1/2 in. above the chalk line (if used) or nail a course of shingles even with the line. Stagger (break) the joints.

Step 5   To cut around doors and windows, position and mark shingles for cutting. Square the lines to form the cutout area. Cut with a crosscut saw and nail in place.

BSid10: CUTTING, NAILING, AND INSTALLING VERTICAL TONGUE-AND-GROOVE SIDING

RESOURCES

Materials:

_____ sq ft tongue-and-groove siding by type.
   no.

Type _____ (unless calculated in Routine BSid1)

_____ lineal ft 1 x 2 or 1 x 3 (circle one) for stripping (if required)
   no.

_____ lb 8d common nails. One pound is enough for 32 sq ft of
   no.

siding.

Tools:
1 jack plane
1 chalk line
1 folding ruler
1 13-oz hammer
1 no. 8 crosscut handsaw
1 pair sawhorses
1 level

1 1/8-in. nail set
1 framing square
1 combination square
1 level
1 brace
1 3/4- to 1-in. augur
1 keyhole saw
1 set wood chisels

ESTIMATED MANHOURS

30 minutes per 4 sq ft

PROCEDURE

Preliminary steps Install building paper (Routine BSid2) and flashing over doors and windows (Routine BSid5).

Step 1 Measure the height of the wall from the sill/foundation to the top of the wall (lookout ledger, frieze, etc.).

Step 2 Precut six or more pieces of siding according to the length measured in step 1.

Step 3 Prepare the starting corner siding board by cutting a groove from the board and dressing with a jack plane.

Step 4 Snap chalk lines (vertically) at 4-ft intervals for guidelines.

Step 5 Install the corner board by face-nailing and keeping dressed edge *flush* with the wall corner. Hold the board level against the edge and adjust for plumb as required.

Step 6 Insert the next board's groove into the first's tongue and toe-nail between the outer surface and tongue on the second board.
Tip: Toe-nail the center, top, and bottom *and* center bottom and top alternately to keep the line plumb.

Step 7 Repeat steps 5 and 6, checking the distances from the board to the chalk line for evenness from time to time.

Step 8 For windows and doors, mark the board for stock to be cut away for the window or door.

Step 9 Lay out the area to be cut away by using a combination square and framing square.

Step 10 Cut stock away:

a. Make cross-cuts.

b. Close to the line drawn, drill a series of holes with the brace and auger on the stock to be removed.

c. Start the saw track by using the keyhole saw.

d. Finish cutting with the handsaw.

e. Even the edges with a sharp wood chisel.

Note: Undercut all cuts made around doors and windows to ease installation.

Step 11 Install a prepared piece of siding and face-nail close to the casing.

Step 12 To close the corner, position the final board and mark by scribing the back edge.

Step 13 Cut along the outside edge of the line drawn. Dress slightly with the plane.

Step 14 Install the board and nail. Finish dressing the edge flush with the other wall siding.

BSid11: INSTALLING VERTICAL-PANEL SIDING

RESOURCES

Materials:

_____ 4 x 8 x _____ panels of type _____, grade
   no.   thickness

_____, manufacturer _____ (unless computed in Routine BSid1). Length of wall divided by 4 ft = no. panels per wall.

_____ lb panel nails. One pound is enough for three panels.
   no.

_____ lineal ft of 1 x 3 stripping (if required). Strips are 16 in. OC
   no.

plus 1 at bottom around perimeter *and* around windows and doors.

Tools:
1 framing square
1 chalk line

1 brace and augers
1 13-oz claw hammer
1 level
1 crosscut handsaw
1 straightedge
1 set wood chisels
1 jack plane
1 pair sawhorses

ESTIMATED MANHOURS

30 minutes per sheet

PROCEDURE

Preliminary steps Install building paper (Routine BSid2) and flashing (Routine BSid5).

Step 1 Measure the full length of the wall and layout the paneling.

a. Divide the wall length by 4 and obtain the number of full sheets plus remainder.

b. *If* the remainder is 6-in. or less, use steps 1c through 1f. If not, skip to step 2.

c. From the center of the span, measure 24 in. each way and mark the wall.

d. Follow by marking 4-ft layouts until the corner is reached.

e. Measure the remaining distance from the last *full*-sheet layout to the corner.

f. Rip a sheet of paneling, keeping the machined edge for butting to the next sheet.

Step 2 Measure, mark, and precut panels for needed height.

Step 3 Install the first panel flush with the outside corner and the inner edge, breaking on a 2 x 4 stud. Nail according to Table 16-2.

Step 4 Lay out and cut out for window or door.

a. Measure the distance from the last sheet installed to the casing edge. Mark on the panel to be cut.

b. Measure from the foundation line to the bottom of the sill and/or window/door drip cap. Mark on the panel.

c. Connect the pencil mark with a straightedge or chalk line and cut away the stock.

d. Install following step 3.

Note: Allow 1/8-in. clearance around all casings.

Step 5 Cut and nail battens and outside corner boards (see Routine BSid3).

Step 6 Cut and install moldings (see Routine BCor4, Soffet Installation), and caulk (see Routine BSid12).

BSid12: CAULKING

RESOURCES

Materials:
_____ tubes caulking. One tube covers approximately 50 ft of joint
   no.

   area.

Tools:
1 caulking gun
1 knife

ESTIMATED MANHOURS

20 minutes per tube

PROCEDURE

Step 1 Insert the tube into the gun.

Step 2 Cut the plastic tip on a 30-degree angle to create a 3/8-in.-diameter hole; puncture the plastic film through the nozzle.

Step 3 Engage the trigger and squeeze.

Step 4 Align the nozzle within the joint area, squeeze the caulking from the gun, and slowly slide the nozzle down the joint, creating a bead.

Step 5 Release the trigger assembly upon completion of the run to stop the flow of caulking from the tube.

UNIT THREE

Interior Installations

The five chapters of Unit Three provide you with descriptions of tasks necessary for completion of interior walls, ceilings, floors, and trim work. Some of the tasks require considerable dedication to accuracy. All of these features will be highly visible, so the joints and splices must be made cleanly and accurately. Sharp tools, accurate marking, and very close fitting are of greatest importance. The illustrations provided will aid you and should answer your questions. When using the routines, be sure to include them in your programmed plan. The code for routines in this unit is I, which indicates that the tasks all relate to interior installations.

17

DRYWALL INSTALLATION

This is the first chapter in a series dealing with tasks performed inside the house. Because of its covering properties; ability to be painted, papered, and texturized; and relatively inexpensive cost, drywall paneling has replaced lath and plaster. As you will see, the routine steps required for a satisfactory installation require some preplanning and careful follow-through.

SCOPE OF THE WORK

As a rule, a helper is needed for the job. This is especially so if ceilings and horizontal sidewall panel installations are planned. The sheets which range from 8 to 14 ft long are not light. Both persons will probably nail while holding the sheets in place. Some scaffolding may be needed if ceilings are installed.

Two professional drywall hangers may do a house in a day, but you had better plan on a room a day (ceiling included). Doors, windows, utility outlets, and corners must be cut and nailing at specific intervals is required. These tasks and others will take considerable time.

Before any drywall job can begin, a close examination of the room to be completed must be made. Each corner must have two nailing surfaces. They may be a ceiling and sidewall corner, or a vertical interior corner. For example, partitions that were installed parallel to the ceiling joist need a cap to form a corner nailing surface (Figure 17-1).

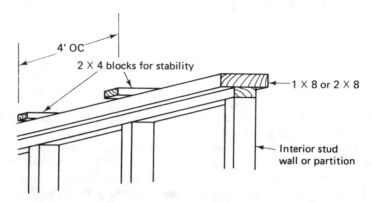

Figure 17-1 Completing Ceiling Corner Framing

FRAMING CORNERS

Notice that either a 1 x 8 or 2 x 8 member is nailed on top of the plate with some exposure on both sides of the plate. Also note that 2 x 4 braces must be cut between joists and nailed through the joists and down into the plate. These 2 x 4 blocks, spaced at an average 4 ft OC prevent the 1 x 8 from lifting when the drywall is nailed in place.

Inside vertical corners not previously completed must be finished. Use Routine BF5, Constructing Inside Corners, in Chapter 7 if you need direction. If areas around utilities such as plumbing need bracing, do this before installing the drywall. If the walls have not yet been insulated, this should be done before the paneling is installed. Follow the manufacturer's suggested instructions. To aid you with the proper selection of installation techniques and material sizes, consider the following plans.

VARIETY OF SIZES OF MATERIAL

Panels are generally 4 ft wide and from 8 to 14 ft long, with 1/2 in. as the usual thickness. Panels 3/8 in. thick may also be used, especially if double-wall panels are installed, or when plywood paneling is to be installed over the drywall. For single drywall application, use 1/2-in. thick panels.

When sheets are installed vertically on a wall, the 8-ft sheet will probably be selected. However, when sheets are installed horizontally on a wall, use the longest sheet possible according to the length of the wall. This will result in only one joint 4 ft from the floor. Minimizing the number of joints should always be of primary importance. Ceilings may also be installed with a minimum of joints by careful selection of lengths.

Along with selecting the size of sheet that is best for the job, you will need to plan to finish the joints and cover the nails. Table 17-1 provides data about the tape and compound necessary for the job. In addition, an average of 6 lb of drywall nails will usually be adequate to nail 1,000 sq ft of panel.

TABLE 17-1: Tape and Nail Requirements

With This Amount of Drywall Panel	Nails Required	USE	This Amount of Powder-Type Compound	OR	This Amount of Ready-Mixed Compound	AND	This Amount of Reinforcing Tape
100 sq ft	0.6 lb		6 lb		1 gal		37 ft
200 sq ft	1.1 lb		12 lb		2 gal		74 ft
300 sq ft	1.6 lb		18 lb		2 gal		111 ft
400 sq ft	2.1 lb		24 lb		3 gal		148 ft
500 sq ft	2.7 lb		30 lb		3 gal		185 ft
600 sq ft	3.2 lb		36 lb		4 gal		222 ft
700 sq ft	3.7 lb		42 lb		5 gal		259 ft
800 sq ft	4.2 lb		48 lb		5 gal		296 ft
900 sq ft	4.8 lb		54 lb		6 gal		333 ft
1,000 sq ft	5.3 lb		60 lb		6 gal		370 ft

Space the nails 7 in. on the ceiling, 8 in. on the wall.

CUTTING TECHNIQUES

Drywall panels are cut to fit with a utility knife or a crosscut hand saw. Where a knife is used, the standard sequence of steps is:

1. Cut through the face paper into the gypsum along a previously marked line.
2. Snap the gypsum by striking a blow behind the cut and fold the two halves to form a 90-degree corner.
3. Cut back the paper with a knife.

Where a saw is used, usually one or more sides need to be cut from an internal angle (does not extend from edge to edge). The saw is used on all cuts or in combination with the knife. The saw is used to cut at least all but one side. The standard sequence for making internal cuts is:

1. Drill a 1-in. pilot hole in each corner.
2. Start the cut with a keyhole saw.
3. Finish the cut with a no. 8 crosscut handsaw.

Panels may be positioned in two convenient positions for cutting (Fig. 17-2): (1) they may be leaned against a wall face side out, or (2) they may be laid flat on a pile of sheets or on a pair of sawhorses.

Figure 17-2 Cutting Panels

NAILING TECHNIQUES

Drywall nails may be purchased in a variety of styles, however, for the 3/8- and 1/2-in.-thick panels, the 1- and 1/4-in. nail with a 1/4-in.-diameter head is used. The two methods of nailing the sheets to framing members are the single and double methods (Figure 17-3).

Because drywall paneling is flexible, a standard method of nailing has been adopted by the trade. Nailing starts at the center of the sheets and spreads to both ends. This method eliminates bulging or twisting.

The recommended nailing techniques require:

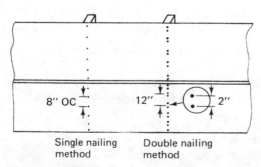

Single nailing method   Double nailing method

Figure 17-3 Nailing Methods for Drywall

1. That nails be driven at least 3/8 in. from the ends or edges of panels.
2. Positioning nails on adjacent ends or edges opposite each other.
3. That you begin nailing from the center of the panel and proceed toward the outer ends or edges.
4. Manual pressure on the panel adjacent to the nail being driven to ensure that the panel is secured tightly to the framing member.
5. That nails be driven with the shank perpendicular to the face of the panel.
6. That the last blow of the hammer seats the nail so that the head is in a slight uniform dimple (Figure 17-4).

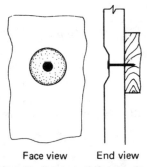

Face view   End view

Figure 17-4 Nail Seated Properly in Dimple

Refer again to Figure 17-3 and note the spacing of nails. Nails in the center of the panel are spaced 12 in. OC along a framing member. The end or edges of panels ending on a framing member have nails spaced 8 in. OC.

METAL CORNER BEAD

Metal corner bead should be installed on each exterior corner covered with drywall. Its function is twofold: (1) it reinforces the corner against wear and tear, and (2) it establishes an excellent reference to finish a corner with compound.

A metal corner is usually made from heavy-gauge hot-dipped galvanized steel. Standard sizes are 15/16 in. x 15/16 in., 1 in. x 1 in., 1 in. x 1-1/4 in., 1-1/8 in. x 1-1/8 in., and 1-1/4 in. x 1-1/4 in. Figure 17-5 shows what a typical corner bead looks like and how it is installed.

After cutting a strip to fit a corner, nail the metal in position with nails spaced 8 in. OC and adjacent. Around window openings metal corners may be cut and bent so that one piece forms both corners at the

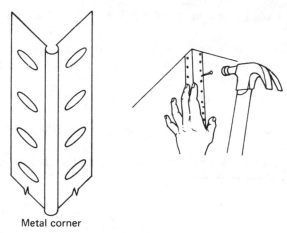

Metal corner

Figure 17-5 Metal Corner Bead   (*Courtesy of United States Gypsum Corp.*)

head. Another technique to use when joining ends of pieces is to use either a 6d or an 8d common nail with its head removed as a pin, fitting one bead to the other. With nailing and corners installed, the finishing steps may begin.

TAPING AND FILLING TECHNIQUES

Butter compound into the channel formed by the tapered edges of the paneling with a broad steel finishing knife. Fill the channel fully and evenly (Figure 17-6A). Center reinforcement tape and press it down into a fresh joint compound. Hold the knife at about a 45-degree angle to the board and draw it along the joint with sufficient pressure to remove the excess compound (Figure 17-6B). Leave sufficient compound under the tape for a proper bond but not over 1/32 in. thickness under the feathered edge.

COVERING TAPE AND NAILS

When the tape is embedded, apply a skim coat of joint compound immediately after embedding (Figure 17-6C). This skim coat reduces the possibility of wrinkling or curling of the edges, which may lead to edge cracking. Allow to dry completely.

Apply the first coat of joint compound over all the fastener heads (Figure 17-6D) immediately prior to or after embedding the tape. Allow the compound over the fastener heads to completely dry.

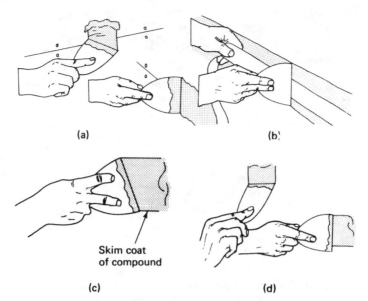

(a) (b)

Skim coat
of compound

(c) (d)

Figure 17-6 Taping and Sealing Panels
(*Courtesy of United States Gypsum Corp.*)

APPLICATION OF SECOND COAT

After the embedding and covering coat is completely dry (24 hours under good drying conditions), apply a second coat, feathered approximately 2 in. beyond the edges of the first coat (Figure 17-7A).

APPLICATION OF THIRD COAT

After the second coat is dry, sand lightly with 220-grit paper. Apply a thin finishing coat to the joints and fastener heads (Figure 17-7B). Feather the joint edges at least 2 in. beyond the second coat. Sand lightly when dry with 220- or 320-grit sandpaper.

FINISHING INSIDE CORNERS

Fold the tape along the center crease. Butter both sides of the corner with joint compound and apply tape (Figure 17-8A). Apply second and third coats of compound (one side at a time) in the same way that you would finish flat joints.

Tape and compound can be applied to both sides in one operation. The corner tool (Figure 17-8B) is angled slightly so that about 1-1/2 in.

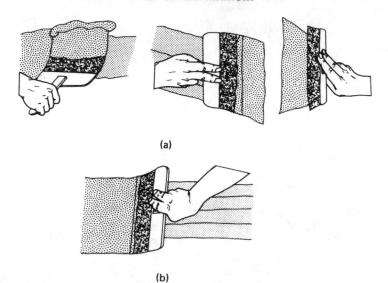

(a)

(b)

Figure 17-7 Application of Second and Third Coats
(*Courtesy of United States Gypsum Corp.*)

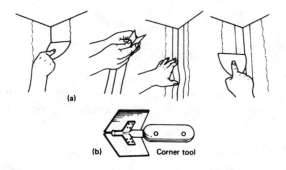

(a)

(b)   Corner tool

Figure 17-8 Finishing Corners   (*Courtesy of United States Gypsum Corp.*)

of blade tip contacts the corner. After embedding the tape, remove the excess compound with the tip of a blade. Final finishing is done with long, continuous strokes.

ROUTINES

Six routines follow. Because there are always variables in framing, the first routine, IW1, provides guidelines for the correction of framing deficiencies. Since there is considerable calculation required in estimating materials, the second routine, IW2, provides a plan for installation and estimation of drywall panels. Routines IW3, Cutting Out for Windows,

Doors, and Utilities; IW4, Cutting, Nailing, and Installing; and IW5, Installing Metal Outside Corners, outline the installation tasks. Routine IW6, Taping and Sealing, outlines the hand method of sealing panels for finishing.

IW1: FRAMING CORRECTIONS

RESOURCES

Materials:
_____ lineal ft of 1 x 8 or 5/4 × 8 (circle one). One piece is needed
no.

for *each* partition that runs parallel to the ceiling joist run.
1 lb 16d common nails per five 2 x 4 blocks
1 lb 8d common nails per 25 lineal ft
_____ 2 x 4 x 16 or 24-in. blocks at 1 per 4 ft over 1 x 8. Con-
no.

verted to total lineal feet: _____.

Tools:
1 no. 8 crosscut handsaw
1 13-oz claw hammer
1 6-ft folding ruler
1 sawhorse
1 combination square

ESTIMATED MANHOURS

1 hour per 12 ft

PROCEDURE

Step 1 Measure and cut a 1 x 8 for each partition that runs parallel to the ceiling joist.

Step 2 Nail the 1 x 8 on *top* of the plate, dividing its width so that some extends over both sides, forming corners.

Step 3 Cut and nail 2 x 4 braces on top of the 1 x 8 and nail into the joists as well as the 1 x 8. Space 2 x 4 blocks at 4-ft intervals.

Note: Complete any interior corners omitted during the framing phase using Routine BF5 as an aid.

IW2: LAYOUT AND ESTIMATING

RESOURCES

Materials:
Floor plan of building
Also see Procedure section

Tools:
1 12-ft tape measure

ESTIMATED MANHOURS

1 to 2 hours

PROCEDURE

Step 1   Select a room to be drywalled.

Step 2   From a measurement or plan, record the width, length, and height of each wall.

wall 1 _____w, _____l, _____h
wall 2 _____w, _____l, _____h
wall 3 _____w, _____l, _____h
wall 4 _____w, _____l, _____h
wall 5 _____w, _____l, _____h

Step 3   From a measurement of the room length or width (step 2), record the ceiling run and width.

ceiling _____r (right angles to ceiling joists)
ceiling _____w

Step 4 Calculate the number of sheets required according to length (all sheets 48 in. wide). Available 8 to 14 ft in length.

_____ **4 ft x 8 ft x 1/2 in.**
   no.

_____ **4 ft x 10 ft x 1/2 in.**
   no.

_____ **4 ft x 12 ft x 1/2 in.**
   no.

_____ **4 ft x 14 ft x 1/2 in.**
   no.

Note: Horizontal wall installations usually result in savings of material and cost.

Step 5 Repeat steps 1 through 4 for each room.

Step 6 Summarize totals and list in programmed plan.

IW3: CUTTING OUT FOR WINDOWS, DOORS, AND UTILITIES

RESOURCES

Materials:
None required

Tools:
1 6-ft folding ruler
1 utility knife
1 48- to 60-in. straightedge
1 keyhole saw
1 brace and 1-in. auger

ESTIMATED MANHOURS

10 minutes per cutout
Note: Do not add this time to the job if Routine IW4 manhours are used.

PROCEDURE

Step 1 From the ceiling, measure to the upper and lower sides of the area to be cut out. Transfer the measurements to a sheet (face side up).

Step 2 From a corner or an edge, or from the end of the sheet, measure the far- and near-side distances of the cutout. Transfer the measurements to a sheet.

Step 3 Connect the measurement points with a pencil and straightedge to form the cutout, except for circular (pipe) cutouts.

Step 4 Check your measurements from the sheet to the wall.

Note: Use steps 5 and 6 for internal cuts; use steps 7 and 8 for external cuts; use step 9 for circle cuts.

Step 5 Drill holes at corners with a brace and auger.

Step 6 With a keyhole saw, cut 4 to 5 in. along the line.

Step 7 Finish cutting with a handsaw, starting from the edges.

Step 8 Cut the paper along a third line by breaking it at the back with a utility knife and cutting back the paper along the crease.

Step 9 Use dividers to draw a circle that connects all four points measured.

Step 10 Drill a pilot hole in the waste-material area with the brace and auger.

Step 11 Cut out the circle with a keyhole saw.

IW4: CUTTING, NAILING, AND INSTALLING PANELS

RESOURCES

Materials:
See Routine IW2

Tools:
1 utility knife
1 13-oz claw hammer

1 48- to 60-in. straightedge
1 keyhole saw
1 no. 8 crosscut handsaw
1 brace and 1-in. auger
1 6-ft ladder
1 scaffold and sawhorses, as required

ESTIMATED MANHOURS

1 person   20 minutes per sheet (walls)
2 people   15 minutes per sheet (ceiling)

PROCEDURE A: CEILING INSTALLATION

Step 1   Measure and cut a sheet of drywall so that its end breaks on a ceiling joist.
a. Mark two places from a common reference point (end of the sheet).
b. Hold the sheet straight on the marks.
c. Cut through the paper with a utility knife.
d. Crack the panel along the cut.
e. From the back side of the panel, cut the paper along the crease.
f. Sand the rough edge, if required.

Step 2   With two people and scaffolds as required, raise the panel to the ceiling, butt to the wall, and corner and nail from the center of the sheet to the ends.

Step 3   Nail to each joist at 8-in. intervals and across the ends at 6-in. intervals.

Step 4   Repeat steps 1 and 2 as required.

PROCEDURE B: HORIZONTAL WALL INSTALLATION

Step 1   Measure and cut a sheet the full length of the wall or to break on a stud. (*Note:* Window and door cutouts may be made before or after installation.)

Cutouts before installation:
1. Measure the distances to the window and/or door from the ceiling and from the wall, corner, or end of the previous sheet. Record on the sheet (face side out).

2. Connect the marks with a straightedge and a pencil.
3. Cut out the waste material:
    (a) Predrill a starter hole on the internal corners.
    (b) Start the cut with a keyhole saw and finish with a handsaw.

Step 2 With a helper, raise the panel to the ceiling and nail in place.

Step 3 Measure and cut the lower piece of paneling (same wall).

Step 4 Mark for the utility outlet and cut out as required. (See Routine IW3, if necessary.)

Step 5 Position the panel against the wall. Slide a flat bar under the panel and raise. Nail in place.

PROCEDURE C: VERTICAL WALL INSTALLATION

Step 1 Measure the panel for the width needed to break on a stud. Cut as required, starting from a corner.

Step 2 Precut all door, window, and utility openings. (See Routine IW3, if necessary.)

Step 3 Position the panel against the wall. Raise the panel to the ceiling with a flat bar. Nail in place.

IW5: INSTALLING METAL OUTSIDE CORNERS

RESOURCES

Materials:
_____ corner beads per 8 ft based on the results of step 1 and 2 of no.

Procedure
1 to 3 lb drywall nails

Tools:
1 pair snips
1 13-oz hammer
1 6-ft ruler
1 stepladder (optional)

ESTIMATED MANHOURS

20 minutes per 8 ft

PROCEDURE

Step 1 Measure and record the lineal feet of each window, door, and outside corner of the drywall installation.

Step 2 Divide the total footage measured in step 1 by 8 and then add one piece to the total. Record your answer in the material portion of this routine.

Step 3 Measure and cut a piece of the corner bead to fit a corner.

Note: Corners around window and door heads and jambs may be neatly formed by bending the metal after determining where the bend will be and only cutting the outside flange.

Step 4 Position the corner bead evenly over the drywall and nail both flanges with nails opposite each other at 6-in. intervals.

IW6: TAPING AND SEALING

RESOURCES

Materials:
1 roll of tape per 60, 200, and 500 lineal ft of joint (select one)
5 gallons mixed joint compound (average)

Tools:
1 4-, 5-, or 6-in. joint-finishing (putty) knife
1 10-in. joint-finishing knife or 10-in. metal trowel
1 stepladder

ESTIMATED MANHOURS

8 to 12 hours per room for three applications

PROCEDURE

Step 1   Fill all cracks and nail dimples with a coat of compound. Wipe off the excess. Let dry.

Step 2   Apply a fair coat of compound to a joint with a 4- or 5-in. knife.

Step 3   Measure and tear a strip of reinforcing tape from a roll as long as the joint.

Step 4   Press the tape into the compound while holding the knife at about a 45-degree angle.

See step 8 for corner installation.

Step 5   Apply another thin coat of compound over the tape to cover the tape, remove the wrinkles, and prevent the edge from shrinking. Let dry.

Step 6   Apply a second coat over the dried first coat, feathering at least 2 in. to each side of the first coat. Let dry.

Step 7   Sand the dried second coat, and apply a thin third coat over the joints and nails with a wide knife. Sand when dry.

Step 8   Crease a strip of reinforcing tape along its center.

Step 9   Press the paper into the previously applied compound, smooth with a knife, and cover with a thin coat of compound.

Step 10   Repeat steps 5 through 7 for corner finishing.

18

CEILING TILE
INSTALLATION

BASIC TERMS

Flat bar 3-in.-wide steel tool representing a stiff putty knife.

Stripping (furring) installing 1 x 3 or 1 x 4 stock across ceiling joists.

Tile ceilings are usually installed *after* the walls have been installed. This differs from drywall ceilings, which are usually installed first. The advantage of the tile block ceiling over the conventional drywall ceiling is that once it is installed, it is finished, whereas the drywall ceiling must be taped and painted.

Regardless of the size or type of ceiling tile being used, its installation is relatively simple and few tools are needed. Various preliminary tasks must be performed to ensure that the finished job is completely satisfactory. Routines are provided at the end of the chapter, as are installation techniques such as stapling and gluing.

SCOPE OF THE WORK

After an examination of the ceiling area and selection of materials and methods, considerable time must be dedicated to preparing the ceiling for the tile. Stripping made from 1 x 3 or 1 x 4 (nominal) stock needs to be installed across open joists. If an old ceiling is being recovered with tile block, some housekeeping chores need to be performed. For the staple method, stripping must be installed, in contrast to the cementing method, where a solid (preinstalled ceiling) surface is needed. Two people will be needed to cut and nail the long strips of 1 x 3 or 1 x 4 to the joists. However, when the tiles are installed, one person can do the job as well as two. Be aware that some form of scaffolding, such as a stepladder or sawhorse and planking, will be needed to make working on the ceiling more comfortable.

Finally, if an old ceiling is to be tiled over, all molding must first be removed, all light fixtures must be dropped, and the heavy ones should be disconnected completely. Be sure to remove the main power to the lights before disconnecting; the wall switch will not, as a rule, remove the power.

One of the first considerations to be examined is the choosing of the size of tiles for the ceiling. Their size will determine, in part, the procedures used for the installation. Each size of course will provide a different overall appearance. The sizes are:

1. Square, 12 in. x 12 in.
2. Square, 16 in. x 16 in.
3. Rectangle, 12 in. x 48 in.

The square tiles are sold with beveled edges and plain edges. The 12 in. x 48 in. tile is sold with plain edges. The beveled edges make for a block pattern across the ceiling, whereas the square or plain edge creates a solid look, as Figure 18-1 shows.

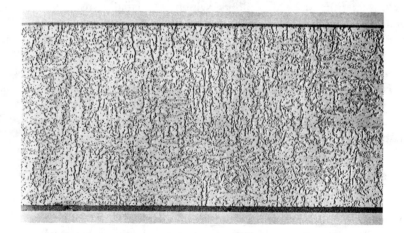

Figure 18-1 Solid Look *(Photo courtesy of Armstrong World Industries)*

The two methods of installing ceiling tiles are the staple method and the cementing method. Because they are significantly different, they are discussed separately in this section.

STAPLE METHOD

If 12- or 16-in.-wide tiles are to be installed, the ceiling, with bare joists or drywalled, must be furred with 1 x 3 or 1 x 4 (nominal) lumber.

These furring strips must be nailed at right angles to the joist run. The spacing on centers must be 12 or 16 in., except for border tiles.

The ceiling has a better appearance if the border tiles are the same width on opposite sides of the room. To calculate where the second strip will be installed to make border tiles equal widths (the first is against the wall), consider the following sample ceiling. The size, as shown in Figure 18-2, shows the short wall at 10 ft 8 in. and the long wall at 12 ft 4 in.

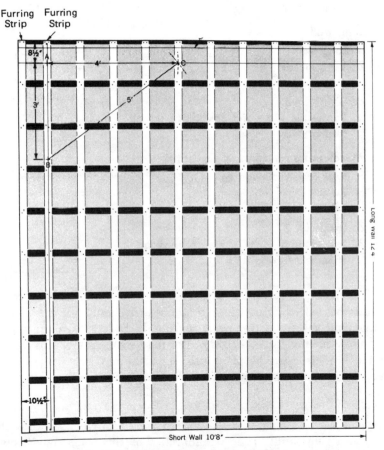

Figure 18-2 Furring the Ceiling

The calculations needed to establish the width of the border tiles and to determine the center position of the second furring strip are shown in the following example.

EXAMPLE:

The lengths of the wall are taken from Figure 18-2.

1. *Short wall*

a. Short wall	=	10 ft 8 in.
b. Extra inches (over even feet)	=	8 in.
c. Add (width of tile block)	+	12 in.
d. Divide by 2		$2\sqrt{20}$ in.
e. Border tile width and center of second furring strip	=	10 in.

2. *Long wall*

a. Long wall	=	12 ft 4 in.
b. Extra inches	=	4 in.
c. Add	+	12 in.
d. Divide by 2		$2\sqrt{16}$ in.
e. Border tile width and position of second furring strip	=	8 in.

3. *Modifications for 16 in. × 16 in. tile*

a. Wall length divided by		16 in.
b. Extra inches	=	?
c. Add		
d. Divide by 2		$2\sqrt{?}$
e. Results equal width of tile and position of second furring strip	=	_____

After nailing the border strips, first and second, each succeeding strip must be nailed at exact on-center spacing until all are installed. Note that end splices of furring strips should be centered on a ceiling joist to allow nailing for both ends. Also note that ends of furring strips along the borders must be firmly nailed. This may require the use of Routine IW1, Framing Corrections, for completion of wall-to-ceiling corners.

After the furring strips are installed, a pair of reference lines must be snapped with a chalk line. These two lines (shown in Figure 18-2) must form a right angle. The irregularities of the walls, bows, out of square, etc., must be accounted for in cutting border tiles to fit.

Cutting and Installing the Tile

Ceiling tiles are easily cut with a sharp handsaw, coping saw, or utility knife. All cuts are made from the top/finished side of the tile.

Since each tile has straight sides, a framing square and/or a combination square can be used to draw cutting lines.

Assume, for example, that the corner tile is to be cut to fit the ceiling shown in Figure 18-2. We see that one dimension must be 10-1/2 in. and the other 8-1/2 in. Lay out the cutting lines. The flanges must remain intact because they are the fastening surfaces for the staple gun. Figure 18-3 shows how the layout is made. After cutting the tile, it is stapled to the second furring strip and border strip, and a nail is driven through the tile into the corner.

The second tile is cut so that three sides, the two with flanges and one grooved, remain intact. It is stapled along line AB and one or two staples are inserted in the border strip. The third tile is cut so that its tongue edge will fit into the corner tile and the flange will fall on line AC. The fourth tile will slip into the opening between the second and third tiles. This process is repeated across the ceiling.

The final border row of tiles is cut to fit the space remaining. A flat bar should be used to draw the tiles together evenly. Face-nail along the border to hold the last corner tile.

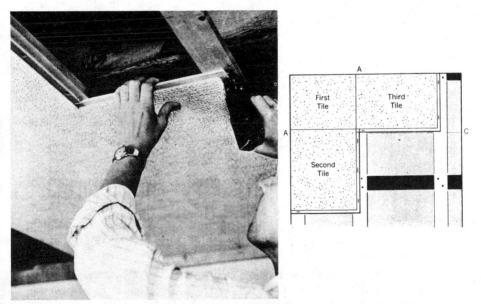

Figure 18-3 Cutting and Installing the Tile

The ceiling will need a molding along the border, and directions for installing it are provided in Chapter 19. But first let's examine the process of cementing a tile ceiling.

CEMENTING METHOD

The cementing method presents a different set of requirements. In brief, the tiles are preconditioned with dabs of acoustic cement and pressed onto a previously installed ceiling such as drywall gypsum board.

There are four basic steps to an installation. They are: (1) preparing the surface, (2) determining the room layout, (3) cutting and installing the ceiling tile, and (4) installing the wall molding.

Preparing the Surface

The ceiling to which the tile is adhered must be *level*, structurally sound, and free of dirt and grease. A straightedge should be slid along the ceiling in several directions to identify hollow areas as well as protrusions. When identified they should be either filled in, if hollow, or tapered even, if protruding.

A ceiling is structurally sound if plaster paint or wallpaper will not pull loose. Perform a test by cementing three to four pieces of tile at various places and allow to dry for 48 hours. Pull the tiles free; if plaster, paint, or paper pulls away from the ceiling with the tile, the ceiling is not good enough to use the cement method.

The ceiling should be prepared as follows:

1. Remove the fixtures
2. Remove molding from the borders
3. Eliminate any protrusions
4. Clean the ceiling (painted) with detergent

Determining Room Layout

Following the cleaning and testing steps, the ceiling layout is made using a chalk line and the squaring method discussed earlier. In this method only, the chalk lines are snapped on the ceiling surface instead of on the furring strip.

Cutting and Installing the Ceiling Tile

Each piece of border tile must be individually measured and cut to fit. After cutting, each piece should be tried for fit. If it fits, brush on or dab five spots of cement approximately 1-1/2 in. square and about the thickness of a nickel (1/8 in.).

Press the prepared tile into place and staple through the flange into the ceiling with 3/8-in. staples. These staples are used to hold the tile while the glue dries. The final piece of tile may be face-nailed close to the corner to aid in keeping its position until the glue dries.

The final step in installing the molding may be made with the aids provided in Chapter 19.

Exceptions

As stated earlier, there are a variety of tile sizes and types. The 16 in. x 16 in. tile may, if the joists fall properly, be stapled directly to the joists. If cementing 16 in. x 16 in. tiles, use nine dabs of cement rather than the five used for the 12-in. tiles.

The 12 in. x 48 in. tile, square edge (no seams), whether stapled or glued, does not follow the same directions as the square tiles. For instance, the border tiles are cut for width but the ends are not. Therefore, the remaining piece of tile, finishing the row, should be used to start the next row. This staggering of end joints strengthens the ceiling. In addition, if cementing tiles to the ceiling, use 14 spots applied with a brush or 10 dabs applied with a putty knife (Figure 18-4).

(a) Applying brush-on ceiling cement

(b) Dabbing acoustic cement

Figure 18-4 Brushing and Dabbing Cement on Tile

ROUTINES

The discussions and suggestions just presented provide a complete overview of two methods of installing ceiling tile. Since there are various sizes and styles, some very easy interpretations to the routines' steps may be required. The routines are designed to provide a logical sequence of steps. Individually they include all the steps needed for each element of the task.

IC1: LAYING OUT 12 IN. x 12 IN. TILE

RESOURCES

Materials:

_____ 12 in. x 12 in. tiles in _____ pattern. The number of
  no.

tiles is to be entered after completing the Procedure section.

_____ gallons glue; or
  no.

1 box 1/2-in. staples

Tools:
1 ruler or tape measure

ESTIMATED MANHOURS

1 hour

PROCEDURE

Step 1　Measure and record each room's length and width.
_____l, _____w

Step 2　Add 12 in. to the number of inches greater than even feet for
short wall.

Step 3　Divide the total number of inches by 2 to give you the *width* of
the border tile for the *long wall*.
Example:

Short wall	=	10 ft　8 in.
Extra inches	=	8 in.
Add	+	12 in.
Divide	$2\sqrt{20}$ in.	
Border tile for long wall	=	10 in.

Step 4　Add 12 in. to the number of inches greater than even feet for
long wall.

Step 5　Divide the total number of inches by 2 to give you the *width* of
the border tile for the *short wall*.

Example:

Long wall	= 12 ft	4 in.
Extra inches	=	4 in.
Add	+	12 in.
Divide	$2\sqrt{16}$ in.	
Border tile for short wall	=	8 in.

Step 6 Multiply the total number of feet plus border pieces for the short wall by the total number of feet plus border pieces for the long wall. Add 5 to 10 extra pieces to account for errors made in installation. Record your answer in the material section of this routine and later in your programmed plan.

Note: Use the above method for other sizes of tile and refer to the chapter for directions.

IC2: INSTALLING FURRING STRIPS

RESOURCES

Materials:

_____ lineal ft <u>1 x 3 or 1 x 4</u> furring strips. The number equals 1
no.   **select one**

for each 12 in. + 1 for the end × the length of the room.
3 lb 8d common nails per 100 sq ft of ceiling space

Tools:
1 ruler
1 chalk line
1 13-oz claw hammer
1 no. 8 crosscut handsaw
1 stepladder and sawhorses

ESTIMATED MANHOURS

3.5 hours per 100 sq ft of ceiling

PROCEDURE

Step 1   Determine the run of ceiling joists by visual observation, sounding, or with a hammer and nail.

Step 2   Cut a furring strip to span the room or end on a joist center with its run at right angles to the joists.

Step 3   Nail a furring strip along the ceiling/wall corner.

Step 4   Using data from Routine IC1, lay out the position of the second furring strip so that its center is equal to the width of the end tile + 1/2 in. Mark the ceiling at both ends of the room one-half the width of the strip either side of the center mark previously made and snap a chalk line to connect the marks.

Step 5   Cut and nail a furring strip alongside the chalk line.

Step 6   Measure 12-in. intervals from the second furring strip's edge across the ceiling at both ends. Snap chalk lines connecting the marks.

Step 7   Cut and nail strips until all are installed.

Step 8   Install a final strip along the ceiling/wall corner to complete the task.

Step 9   Box any protrusions with 2 x 4 and 1 x 4 framing materials so that the 12-in.-wide blocks can be fastened.

IC3: MARKING FOR A 3-4-5 RIGHT ANGLE

RESOURCES

Materials:
None required

Tools:
1 6-ft ruler
1 chalk line

ESTIMATED MANHOURS

30 minutes

PROCEDURE

Step 1   Using data from Routine IC1, measure two places on the ceiling or furring strip a distance equal to the width of the first tile and snap a chalk line.

Step 2   Using data from Routine IC1, measure in from the wall along the chalk line a distance equal to the width of its first tile. Mark point A. (See Figure 18-2.)

Step 3   From point A, measure in exactly 3 ft along the chalk line. Mark point B.

Step 4   Starting at point A, measure off exactly 4 ft along the adjacent wall. Mark a small arc.

Step 5   From point B, measure exactly 5 ft toward the 4-ft arc made in step 4. Mark the intersect point C.

Step 6   Snap a chalk line from point A over point C to the opposite wall.

IC4: CUTTING AND STAPLING TILE

RESOURCES

Materials:
Calculated in Routine IC1

Tools:
1 no. 9 or no. 10 crosscut handsaw, coping saw, or utility knife
1 staple gun and staples
1 6-ft folding ruler
1 framing square
1 6-ft stepladder (optional)
1 worktable or sawhorses (optional)

ESTIMATED MANHOURS

3 hours for an 8 ft x 8 ft ceiling

PROCEDURE

Step 1 Cut the first tile so that it fits into the corner. In the example, the first tile would be cut 8-1/2 in. by 10-1/2 in. After aligning the flanges with reference line AB and line AC, staple this tile into position (see Figure 18-3A and B).

Step 2 Cut a second tile so that one of the tongue edges fits into the corner tile and the stapling flange falls directly on line AB.

Step 3 Cut a third tile so that the tongue edge will fit into the corner tile and the stapling flange will fall directly on line AC (see Figure 18-3B).

Step 4 Work across the ceiling, installing about two tiles at a time along the borders and filling in between with full-sized tiles. Staple each tile with three 9/16-in. staples in the flange into the furring strip and one in the flange closest to the wall or previously installed tile.

Step 5 When you reach the opposite wall, measure each border tile individually. Fasten the tile in place by stapling into the remaining flange.

IC5: CEMENTING TILES TO A CEILING

RESOURCES

Materials:
1 gallon of cement (approximately) will be needed for each 100 12 in. x 12 in. tile

Tools:
1 no. 10 crosscut handsaw, coping saw, or utility knife
1 folding ruler
1 1-1/2-in. putty knife
1 chalk line
1 stepladder (optional)
1 workbench or sawhorse (optional)

ESTIMATED MANHOURS

12 tiles per hour

PROCEDURE

Step 1   Clean the old ceiling (if required).

Step 2   Inspect the ceiling and remove any protrusions.

Step 3   Lay out the ceiling according to Routines IC1 and IC3.

Step 4   Cut a corner tile and two adjacent tiles as per steps 2 through 5 of Routine IC4.

Step 5   Put five dabs of cement on the back of the tile—one in each corner and one in the center.

Note: Use 10 dabs of cement on 12 in. x 48 in. tiles.

Step 6   Press the tile firmly to the ceiling and, when aligned, staple with two or three 3/8-in. staples.

Step 7   Continue installing border tiles and full tiles until the ceiling is complete.

19

TRIM

BASIC TERMS

Casing trim stock for the doors and windows.

Coping a method of fitting molding by coping out some stock to make an interior corner.

Molding stock specifically shaped to be used as trim, such as cove, bed, crown, and shoe.

Trimming patterns are very old and are used by the best craftsmen because they always produce the best results. No phase of construction is seen more than the quality of the trimming job. Each joint successfully completed receives more praise for "a job well done" than a poor joint. Be sure, patient, and accurate with your work, and if it's wrong, *do it over*.

SCOPE OF THE WORK

Generally one person can install all the trim in a room without help. However, when installing long strips of ceiling molding, a helper is very useful. Because of the accuracy required, the work is time consuming. It is possible to spend a full day trimming an average-sized room containing a door, window, floor, and ceiling molding. Most of the work can be performed while standing on the floor, but installation of ceiling molding requires a stepladder, bench, or sawhorse.

Finally, but of equal importance, proper planning as to sequence of installation and sizes (lengths) of stock to buy can result in a savings in time and in the cost of supplies.

Figure 19-1 provides an overview of the typical trim and molding used in a home. Each type is identified by name and use. Each type is usually available in several sizes—for example, cove molding may be as small as 3/4 in. or as large as 2 in.

Trim for casings and baseboards usually is planned where the center area of its back edge is hollow. This technique is provided to ensure fitting to the wall surface even where slight irregularities exist

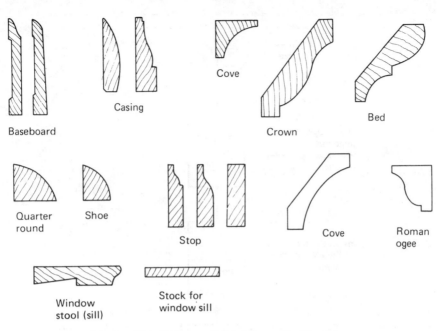

Figure 19-1 Trim and Moldings for Interiors

on the wall. Ceiling moldings, such as cove, crown, and bed, are also designed with space for irregularities behind the molding.

PATTERN OF INSTALLATION OF TRIM

Two patterns of trim installation have been developed in the trade: (1) the entire-room approach, and (2) the individual-element approach.

The entire-room approach includes planning which parts of the room are to be done first, second, third, and so on. The number of items that a room may have which require trim includes:

1. Inside and outside corners
2. Doors to trim
3. Windows to trim
4. Baseboard and shoe molding
5. Ceiling molding

Figure 19-2 is a flow chart which shows the sequence needed for successful installation of trim in a room. To use the chart, simply identify which elements in your room require trim. Note the position of each in the chart and start with the task *closest* to "Start Here" on the chart. Observe the preconditions and determine their applicability and com-

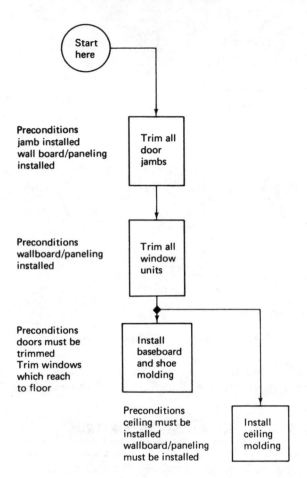

Figure 19-2 Pattern to Trim a Room

pleteness. With the use of the overall pattern shown in Figure 19-2, the individual patterns are developed in sequence.

PATTERN FOR INSTALLATION OF DOOR CASING

There is a precise method to use for the installation of casing trim on a door. Follow the approved method, for two reasons: (1) it provides the most economical use of material with the fewest possibilities for error, and (2) repetition will build confidence and skill.

Figure 19-3 shows the pattern for trimming a door. Left to the choice of the builder is whether to start with the left or right side of

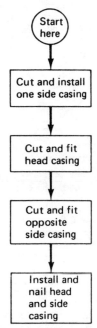

Figure 19-3 Pattern to Install a Door Casing

the door jamb. Install the trim in the sequence that results in the best job.

Note from Figure 19-3 that the side casing is cut, fit, and nailed. Next, the head piece is cut to fit. Do not install the head casing because it will interfere with fitting the other side casing. When the opposite (second) side casing is cut, trial fitting of head and side casing is needed. All nails must be set for filling.

The primary reason that all door jambs must be cased before other work is performed is that their trim rests on the floor and may, in unusual cases, reach the ceiling. Refer to Figure 19-1 and see that trimming windows is the second phase of the task.

PATTERN FOR INSTALLATION OF WINDOW TRIM

As with trimming door jambs, windows are trimmed in a precise sequence. Almost without exception, a window's trim consists of a sill, apron, jamb/boxing, casing, and stop. If wooden window units are installed, the jambs form a part of the window unit. However, aluminum window units usually require formation of a jamb or "boxing" to cover the exposed studs and headers. Figure 19-4 shows in flowchart fashion

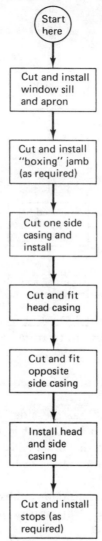

Figure 19-4 Pattern to Install Window Trim

the approved method for installing window trim. The sill is always installed first, followed by the installation of the apron (casing under the sill). The apron provides two functions: (1) it aids in supporting the sill, and (2) it covers the joint below the sill and the wallboard or paneling.

If "boxing" is required, this task would be done next, but note that the chart indicates "(as required)." If boxing is not needed, you would proceed with cutting and fitting the casing. Finally, if required, window stops are cut, fitted, and nailed in place.

A few of the significant reasons why it is necessary to trim windows before installing baseboards or ceiling molding are:

1. Some types of windows reach from floor to ceiling, or one of the ends may be either place.
2. Some windows may have cornice permanently installed, and ceiling molding will be cut around the cornice.
3. You are working with casing when trimming door jambs and should finish cutting and installing casing while the technique is fresh in your mind.

PATTERN FOR INSTALLATION OF BASEBOARD AND SHOE MOLDING

Figure 19-5 illustrates the sequence used for installing baseboard and shoe molding. By starting at a door and installing the pieces in sequence either *to the right* or *to the left*, you will avoid making interior end cuts on both ends of the last piece. Most baseboard is *molded* (not square, as a 1 x 4); therefore, interior corners must be cut and fit using the coping method. Simple 45-degree cuts will not be satisfactory.

All splicing and outside corners must be made using the 45-degree cut. The splice will consist of an open and a closed miter. The outside corner will consist of two closed miters joining at the corner.

After the baseboard has been completely installed, the shoe may be installed. This molding, which resembles quarter-round molding, is installed everywhere there is baseboard. Once again, start at a door and work your way around the room.

PATTERN FOR INSTALLATION OF CEILING MOLDING

The pattern for installing ceiling molding is essentially the same as for baseboard installation. As Figure 19-6 shows, the pieces are cut to fit an interior corner. If the molding is long enough, cut it to reach the opposite corner. *Coping* is the method used for making corner molding fit. The last wall to receive the molding should be either a short wall, to make fitting a single length of molding easier, or one of the longest, as a splice in molding will make fitting the corners easier. Remember that for the last wall, *both* ends of the molding will need coping. If there is an outside corner in the room, plan the last two pieces of molding to meet

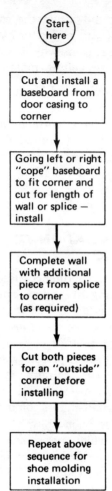

Figure 19-5 Pattern to Install Baseboard and Shoe Molding

at the corner. This will also eliminate the need to make two coped ends on a single piece.

Up to this point, the information and discussion has been about establishing the job and understanding the various requirements. The patterns for installation have been provided and explained. At the end of the chapter are nine routines that provide aids and directions for the installation of every piece of trim required in a room. But before using a routine, let's examine how to prepare each type of cut required in the variety of moldings and trim associated with the numerous tasks.

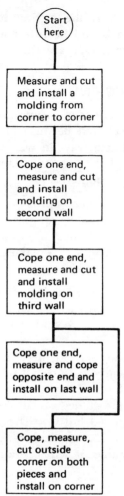

Figure 19-6 Pattern to Install Ceiling Molding

PREPARING FOR EACH TYPE OF CUT

Essentially, except for unusual cases, a square or 45-degree cut is required of every piece of trim. The casings standing on floors or window sills will use the square cut, as will baseboards butting to a door casing or wall. On the other hand, joining of trim at a right angle requires a 45-degree cut.

Because of the improbability of an accurately formed corner at 90 degrees, a special technique for fitting trim and molding is used. This technique, as you suspect, is the coping technique. It is used on a variety of different materials and in different ways. However, the principal sequences of making a *cope* are the same.

TECHNIQUE OF COPING

Three different applications of coping are provided by the examples below. They have been selected because of their applicability to trimming a room and the different ways in which molding and trim are cut. From the examples it will be easy to adapt to any variations or applications that you may encounter.

Example A: Coping for Baseboards

To aid with the explanation, Figure 19-7 is provided. Reading about the task and seeing it as illustrated are not enough. You must, if you have never done it before, practice making a few corners to appreciate the task's requirements.

1. Assume that one piece of baseboard is installed (detail A). Note its butt cut in the corner.
2. The object is to cut a second piece of contour with the first piece's surface.
3. Insert the end of the second piece of stock in the miter box and saw an *open* 45-degree cut (detail B).
4. Lay the trim flat on a sawhorse or bench and *cope* the end stock away from the 45-degree cut. Meticulously follow the edge of the face surface (detail C). The cut completes the task.

   Note: On baseboard and similar trim, *undercutting* the cope is very desirable. It will guarantee a fit even if the wall is out of line.

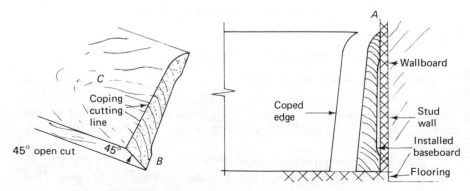

Figure 19-7 Coping Baseboard Trim

5. Place the trim in position and try for fit. If you are sure of the accuracy of the cope and the fit is poor, consider the following possible causes.

   a. The installed piece is twisted because of debris behind the trim.

   b. The floor is uneven and forces the trim up or down, opening the joint.

   c. The corner is not an accurate 90 degrees and forces an open joint.

   d. The coped trim does not have sufficient undercut.

Example B: Coping for Stops and Similar Trim

The head piece of the door or window stop is cut with both ends square. This requires that the side pieces be *coped* to fit the head. Figure 19-8 illustrates the two most common designs used in stops and the coping technique.

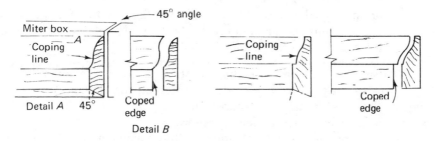

Figure 19-8 Coping Stops

The task requirements are similiar to those of baseboards, but easier.

   1. Insert the molding in a miter box with the molded edge *up* and *facing out.* Saw an approximate 45-degree miter.

   2. With a coping saw (detail A), cut away the miter surface, undercutting by 1 to 3 degrees.

With stops there will be very little need to worry about fit. What must be kept foremost in mind, however, is the need to *cut stops as a pair,* one left and one right.

Note: Although not a part of the coping task, another trick of the trade is to cut stops for length *after* the coping is completed.

Example C: Coping Ceiling Molding

Now comes the difficult task. Coping ceiling trim, such as cove, bed, or crown molding to fit interior corners, requires not only practice but a degree of imagination. If you can grasp, understand, and retain the principles of the following description in conjunction with the illustrations, you can install ceiling molding.

Principles (see Figure 19-9):

1. Ceiling molding is made to be in contact with both ceiling and wall. Therefore, every miter at an interior or outside corner is actually a compound miter (detail A).
2. The edge of the molding that will be in contact with the *ceiling* must be the edge set accurately on the *base* of the miter box. This means that you must insert the molding into the miter box at an angle of 180 degrees from its ceiling position (A of detail B).
3. The edge of the molding that will be in contact with the wall must be the edge set accurately against the *back side* of the miter box (B of detail B).
4. The 45-degree miter that will be cut must result in an open miter with the leading edge of the miter representing the longest part of the coped edge. This means that the length of a piece, if needed, must be marked on the *bottom edge* of the molding as it is installed on the ceiling (detail C).

To summarize, ceiling molding must be inserted in a miter box in a position 180 degrees to its intended position on the ceiling. This is the only method that results in accurate mitering of corners. The cut for coping must be an open 45-degree miter.

Coping requires strict attention to the following details (Figure 19-10):

1. The bottom edge (the edge seen) should be undercut *very slightly* not more than 1 to 3 degrees.
2. The rest of the molding to be coped must be *severely undercut.* This is essential to a good fit.

Figure 19-10 shows how important the undercut is to a good fit. The piece of trim, although cut as a 45-degree miter, is actually a compound miter. It is this combination of angles that makes the undercut needed and important.

TECHNIQUE OF CUTTING A WINDOW SILL AND APRON

Window sills are made from a molded stock called *window sill* or *stool* or from standard nominal 1 x 6 or 1 x 8 stock. The method of

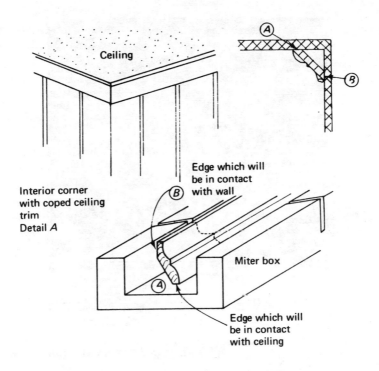

Ceiling

Edge which will be in contact with wall

Interior corner with coped ceiling trim
Detail A

Miter box

Edge which will be in contact with ceiling

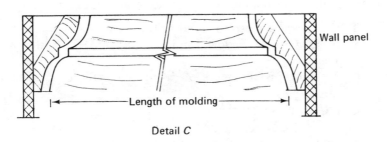

Wall panel

|← Length of molding →|

Detail C

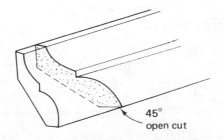

45°
open cut

Figure 19-9 Cutting Ceiling Molding for Coping

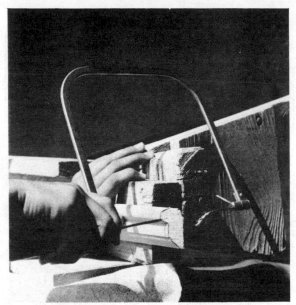

Figure 19-10 Coping Ceiling Molding

making a sill fit an opening is the same regardless of the type of material used. Several significant requirements must be met when laying out and cutting sills. The requirements are as follows (Figure 19-11):

1. The sill's *width* must be equal to the
   a. distance from window frame (aluminum) or sash (wood), plus
   b. the thickness of a piece of casing that will be installed as an apron, plus
   c. a 3/4- to 1-in. overhang.
2. The sill's *length* must be equal to the
   a. length between window jamb (wood) or aluminum window frame, plus
   b. two times the width of the casing, plus
   c. two times an overhang of 1/2 to 1 in.

The sequence of steps required to fit a sill are well defined in Routine IT6, Installing Window Sill and Apron. Briefly stated, a cutout must be made at each end of the sill after its length and width have been prepared. This cutout, as shown in Figure 19-11, allows for: (1) the sill to butt to the window, and (2) the overhang to act as a base for the window casing.

The usual nailing places for securing a sill are at the ends, where the sill passes the jamb, and in the center. In each case the nails will

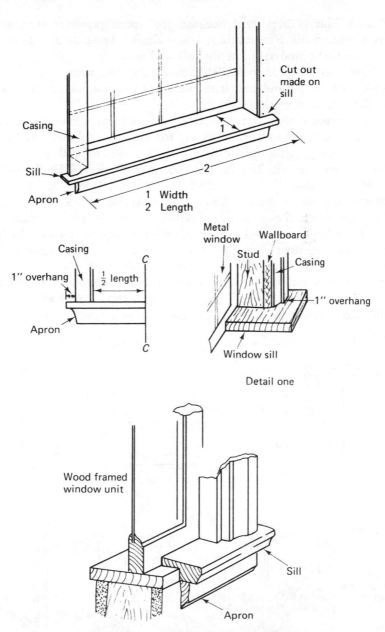

Figure 19-11 Window-Sill Installation

need to be *toed* slightly—the ones at the end, to make driving them easier, and the center, to pull the sill in to the window and down.

The window apron should be installed immediately after the sill is secured. This is important because the apron provides support for the overhanging sill. Essentially it keeps the sill level. It also hides the gap between the underside of the sill and wall.

How does an apron look best? Some carpenters square-cut the ends of a piece of casing and nail it in place. The better job involves a bit more planning and work.

First, most casing is not square and uniform through its width. Generally it is contoured or styled, as, for example, colonial. These distinctive shapes provide you with an opportunity to add style to the aprons you may install. The idea is shown in Figure 19-12. By cutting the contour or shape at the end of the apron, a pleasant break and finishing appearance is made. This type of cut is relatively simple to make.

Figure 19-12 shows a combination square in place against the thick side of the casing and a short block of casing held (with its back to the blade) against the square. A pencil line is carefully drawn along the contoured face. This line is the cutting line.

To make a real professional marking block, spend a few minutes and cope each end of the marking block. This coping will allow you to seat the block exactly onto the contour and make marking very easy.

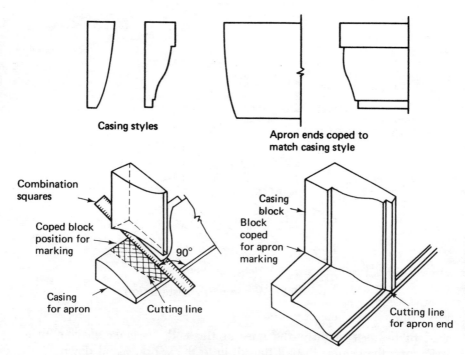

Figure 19-12 Cutting an Apron

TECHNIQUES FOR CUTTING WINDOW OR DOOR CASING

Door and window casing (trim) is usually designed to give a picture-frame effect when installed. Therefore, the thinnest part of the stock is usually nailed to the jamb, and the thickest is nailed to the wall.

Because of this general sloping trend, the top surface cannot be used as a guide for marking 45-degree cuts. That is, the combination square's blade cannot lie flat on the surface while marking for the cut. If you did mark it so, the resultant cut would be somewhat less than 45 degrees, and when the two pieces that make a corner are installed, only the outside edge's corner will touch.

There are two successful ways to cut casing: one is to use a miter box and the other is to develop a technique of holding the square at a level plane and marking the cut accordingly.

The disadvantage of using a miter box is that the accuracy of the cut prohibits undercutting, which is desirable. However, this can easily be overcome by planing an undercut with one or two passes of a block plane.

Cutting casing that has been marked with a combination square provides a feeling of freedom. With it you can adjust your cutting pattern to the trends established during the installation. This marking and cutting method also provides the opportunity to undercut the casing with the initial cut.

The pattern for cutting and installing casing on door units and windows is the same (Figure 19-13). You will develop a sequence of left to right or right to left because one will feel comfortable and will be most successful. The vertical piece should always be cut and installed first. Then both head and opposite side pieces should be cut and fitted before nailing. This method allows for the individual fitting of a corner before proceeding with the next corner. The procedures provided with Routine IT2, Installing Door Casing/Trim, and IT8, Installing Casing on a Window, provide the details for this task.

BOXING IN A WINDOW UNIT

Almost without exception aluminum window units require a boxing effort as part of the interior trim. If you recall, these units are nailed to the outer wall's sheathing and space is provided on each side to allow for a plumb and level installation.

The sill will, in most cases, take up the difference between the sill of framing (2 x 4s) and the bottom of window unit. However, because

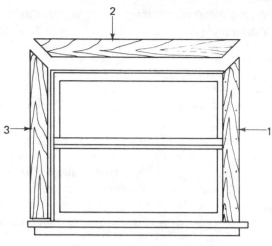

Sequence of casing installation

Figure 19-13 Installing Casing

there may be exceptions, the blocking technique described (following) is applicable under sills as well.

The opening between the framing and the head and sides cannot exceed the thickness of the boxing stock. Let's use an example as shown in Figure 19-14. In our example the distance between side stud and window frame is 1 in. The stock used to box in the window is nominal

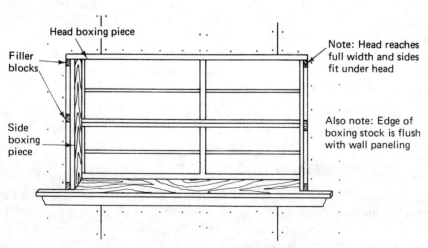

Figure 19-14 Boxing In a Metal Window

1 in. whose actual thickness is 3/4 in. thick. Add to these facts that the box-in stock must overlap the window frame by 1/4 in.

With these facts we can calculate that a filler block is needed. It would be:

1. Distance between stud and window = 1 in.
2. Overlap needed on window = 1/4 in.
3. Total dimension = 1-1/4 in.
4. Less actual stock thickness − 3/4 in.

5. Thickness of blocking stock needed = 1/2 in.

If by chance the stud or header is not parallel to the windows frame, tailored pieces of filler will need to be cut and nailed in place before the boxing pieces of jamb can be cut and fit.

The approved method for boxing in a window is to work on the sill first, then the head, and finally the sides. All cuts should be square with very accurate fitting and very slight undercutting. All pieces should be ripped for the correct width prior to fitting for length. Should the width of the boxing stock be different at the top and bottom or side to side, cut your stock for the widest piece and dress down the narrow end with a jack plane.

ROUTINES

Trimming is the most exacting of all carpentry work. To be good requires not only a comprehensive knowledge of how to do the task, but skill coming from the experience of frequently preforming the various tasks. The information is provided in this chapter, and the tasks are provided in the routines that follow. You must provide the skill by practicing what is written.

IT1: COPING MOLDING FOR INTERIOR CORNERS

RESOURCES

Materials:
Trim and/or molding as detailed in other routines

Tools:
1 commercial miter box and saw, or wooden miter box and no. 10 cross-cut handsaw
1 coping saw
1 set of extra blades for coping saw
1 sawhorse or workbench

ESTIMATED MANHOURS

8 minutes per piece

PROCEDURE

Step 1   Insert the molding or trim into the miter box with the piece-face surface toward you.

Step 2   Position the stock near the end or in line with the mark and required 45-degree cut in miter box.

Step 3   Make an *open* 45-degree cut.

Step 4   Remove the stock and lay flat on a sawhorse or bench.

Step 5   Using the line created by the face side and 45-degree end stock, cut along the line with a coping saw.
a. On *trim,* undercut 2 to 3 degrees.
b. On cove, crown, or *bed molding,* undercut as much as possible, up to 45 degrees where molding is used on ceiling borders.

IT2: INSTALLING DOOR CASING/TRIM

RESOURCES

Materials:
_____ lineal ft trim _____ and _____. Where 17 to 18 ft
　　　no.　　　　　　　　　type　　　　　　size

of trim will cover one side of a jamb with doors up to 36 in. wide. Multiply times the number of sides of doors to trim.

22 to 25 4d finishing nails per door side
22 to 25 8d finishing nails per door side

Tools:
1 ruler
1 no. 10 crosscut handsaw
1 block plane
1 combination square
1 13-oz claw hammer
1 1/16-in. nail set
1 pair sawhorses

ESTIMATED MANHOURS

45 minutes per door side

PROCEDURE

Step 1 Set the combination square for 1/4 in. and scribe a mark 1/4 in. back from the inside of the jamb (side and head).

Step 2 Square and cut one end of a piece of trim.

Step 3 Stand the trim on the floor in position along the pencil line or measure from the floor and mark the distance of the 1/4-in. mark made on the head.

Step 4 Using the 45-degree section of the combination square, mark the trim with the short point being marked equal to the length measured in step 3.

Step 5 Cut the trim along the pencil line (do *not* cut the line away) with the saw, undercutting slightly.

Step 6 Position the inside edge of the trim along the pencil line on the jamb and *tack-nail* 4 in. from the floor, in the center, and 4 in. from the top of the trim. Use 4d finishing nails in the jamb and 8d finishing nails in the wall.

Step 7 Mark and cut another piece of trim to complete the corner started with the first trim.

Step 8 Position the trim in place along the line drawn on the head. (Cut excessively long trim stock to ease handling.)

Step 9 Examine the joint for fit. If slightly open, plane with the block plane until fit. If very poor, recut the head piece accordingly.

Step 10   When fit, position the head trim and mark the opposite end where the vertical pencil line on the jamb intersects the head pencil mark. Cut a 45-degree corner on the head piece.

Step 11   Square an end on a piece of trim and mark for length and cut as in steps 2 through 5 for the other vertical member.

Step 12   Tack-nail the head and opposite vertical pieces of trim in place. Check corners for fit. Adjust if necessary.

Step 13   Complete nailing the trim to the jamb and wall. Drive a 3d or 4d finishing nail at each corner while holding (if required) the outer edges of the trim even. Drive the nail down from the top.

IT3: INSTALLING BASEBOARD

RESOURCES

Materials:

_____ lineal ft baseboard _____ and _____. The num-
        no.   type   size
ber of lineal feet equals the perimeter of the room in length + 4 ft for waste, or the number of lineal feet should be designated in the number of pieces of a specified length to avoid splices.

1 lb 3d finishing nails per job
1 lb 8d finishing nails per 60 lineal ft of baseboard

Tools:

1 6-ft folding ruler
1 adjustable scribe
1 combination square
1 coping saw
1 no. 10 crosscut handsaw
1 13-oz claw hammer
1 1/16-in. nail set
1 pair sawhorses
1 miter box, professional type (optional)

ESTIMATED MANHOURS

16 minutes per 12 lineal ft

PROCEDURE

Step 1 Trim square one end of a baseboard.

Step 2 Measure the length of baseboard needed from corner to corner or door trim to corner, cut, and install with 8d nails.

Step 3 Scribe the next baseboard to the end of the first piece installed, cut or precut an open 45-degree cut at the end of another base-board, and cope along the face line.

Step 4 Measure the length of baseboard needed from the floor level of the first board to the next wall or door trim (unless a splice is required). Square and cut.

Step 5 For the splice, cut an open 45-degree cut on the baseboard being installed.

Step 6 Nail the baseboard in place (glue the joint if desired) on each stud and into the sole/plate.

Step 7 To close the splice, trim another end of a baseboard and fit its end to the corner or door trim.

Step 8 Overlap the open splice and mark the *long* point on the board to be fitted. Mark the baseboard for a *closed* 45-degree cut. Cut the board and try for fit. Nail after adjustments are made.

Step 9 Complete an inside corner fit on a baseboard whose other corner will form an outside corner.

Step 10 Position the baseboard to the inside corner and mark for an outside 45-degree corner. Mark the board for a closed 45-degree corner and cut. Set aside.

Step 11 Repeat steps 9 and 10 for the adjacent side of the outside corner.

Step 12 Nail both pieces in place after making the necessary corrections.

Step 13 Repeat the above procedures until the room is complete. Set all nails with the nail set.

IT4: INSTALLING SHOE MOLDING

RESOURCES

Materials:

_____ lineal ft of shoe molding. The number of lineal feet is equal
  no.

to a room's perimeter + 4 ft for waste.

1 lb 4d finishing nails per 100 lineal ft of shoe molding

Tools:

1 miter box
1 no. 10 handsaw or straight backsaw
1 coping saw
1 13-oz claw hammer
1 1/16-in. nail set
1 6-ft folding ruler
1 sawhorse

ESTIMATED MANHOURS

16 ft per 15 minutes

PROCEDURE

Step 1 Pick a starting point and cut a piece of shoe to fit from a door
to the wall.

Step 2 Cut the end of the shoe near the door trim at an *open* 45-
degree cut. Nail in place with 4d nails at 12 in. OC.

Step 3 Precut an open 45-degree cut in the second shoe. Cope cut (see
Routine IT1, if necessary).

Step 4 Measure from the floor and the installed shoe to the other wall;
or, for a splice, mark the shoe for length and cut. For the *wall,* cut the
shoe square; for a splice, make an open 45-degree cut.

Step 5 Fit the coped area and nail the shoe in place.

Step 6 To complete the splice, butt another piece of shoe to the opposite wall along the baseboard.

Step 7 Mark the piece for a splice cut at the point where the installed open 45-degree cut ends against the baseboard.

Step 8 Insert the shoe in the miter box, align the mark to the slot, and make a *closed* 45-degree cut.

Step 9 Install and set nails.

Step 10 Continue around the room, installing pieces as required.

Step 11 To complete the outside corner, cut, cope (as required), and fit the shoe for the inside corner.

Step 12 Position the shoe on the floor in place and mark the outside corner on top of the shoe.

Step 13 Insert the shoe in the miter box, align the mark on the shoe with the slot in the miter box, and make a *closed* 45-degree cut. Set aside.

Step 14 Repeat steps 11, 12, and 13 for the other piece needed to complete the outside corner.

Step 15 Install both pieces, correcting any errors in fit as needed.

IT5: INSTALLING CEILING-BORDER MOLDING

RESOURCES

Materials:
_____ lineal ft _____ molding. The number of feet
    no.   size and type

equals the perimeter room length plus 4 ft, or the lengths of pieces per room, e.g., two 12-ft and two 13-ft 1-1/2-in. cove.

1 lb 8d finishing nails per 60 lineal ft

1/4 lb 4d finishing nails for outside corners

1 pt white liquid glue (optional)

Tools:
1 miter box

1 no. 10 crosscut handsaw or straight-back saw

1 coping saw

1 13-oz hammer
1 1/16-in. nail set
1 pair sawhorses
1 stepladder (optional)

ESTIMATED MANHOURS

20 minutes per piece with coping

PROCEDURE

Step 1 Measure the length of molding from wall to wall. Cut and nail. *Note:* Be sure that both surfaces of the molding lie flat on the wall and ceiling, respectively.

Step 2 Nail the molding with 6 or 8d finishing nails, aiming the nails slightly up toward the wall plate.

Step 3 Insert the second and remaining pieces of molding in the miter box in a position 180 degrees from its intended ceiling position with the end to be a 45-degree cut near the slot in the miter box. Make an *open* 45-degree cut.

Step 4 Cope molding (see Routine IT1, if needed).

Step 5 Measure the length of piece needed from the point of installed molding *closest* to the wall (bottom point) to the opposite wall.

Step 6 Mark the length measured in step 5 from the *longest* point of the coped end to the opposite end.

Step 7 Cut a piece approximately 1/2 in. *longer* than marked.

Step 8 Try the coped molding for fit to the piece installed. Adjust the cut with a coping saw, if required. If badly fit:
a. Check for proper position of the first piece of molding installed.
b. Check that the cope is undercut sufficiently to allow the corner to mate.
c. Carefully position the molding in the miter box and remake an open
   45-degree cut; then cope again.

Step 9 When fit, reaffirm the length needed and cut square the end of the molding. Nail in place.

Step 10 Cut and fit the interior corner and mark the short point (bottom) of the molding even with the corner.

Step 11 Insert the molding in the miter box 180 degrees from the ceiling position and make a *closed* 45-degree cut. Set aside.

Step 12   Cut another piece from the adjacent wall.

Step 13   Tack one piece and try a second for fit. If the fit is good, glue the joint and nail the molding to the wall. If the fit is poor, recut one or the other pieces.

Step 14   The first step in making a splice is to fit the molding to make an interior corner. Make an *open* 45-degree cut on the opposite end and nail the molding in place.

Step 15   Fit a second piece to the opposite interior corner. Measure the length from the *lower* edge of the installed molding on the adjacent wall to the bottom edge (face surface) of the molding installed in step 14.

Step 16   Insert the molding in the miter box oriented at 180 degrees to the ceiling. Position the length mark so that the saw aligns with the mark when it is over the molding. Cut the molding at a *closed* 45 degrees.

Step 17   Trial-cut a piece for fit both at the internal corner and at the splice. If correct, nail.

IT6: INSTALLING WINDOW SILL AND APRON

RESOURCES

Materials:
_____ lineal ft of sill material of either standard sill or 1 x 6 stock
   no.

(calculate apron with trim in Routine IT8).
6 to 10 6d or 8d finishing nails per window

Tools:
1 6-ft folding ruler
1 combination square
1 block plane and/or
1 12-in. jack plane
1 pair sawhorses
1 1/16-in. nail set
1 coping saw
1 no. 8 crosscut handsaw

1 13-oz. claw hammer
1 no. 5-1/2 ripsaw or power saw (optional)

ESTIMATED MANHOURS

30 minutes per sill and apron

PROCEDURE

Step 1 Prepare the stock for width by measuring the following:
width from window sash or frame to wall
> surface   _____ in.

thickness of trim   _____ in.

extension to room   1   in.

> total width   _____ in.

Step 2 Calculate the length of the sill by measuring the following:
width of window opening   _____ in.

2 × width of window trim   _____ in.

overhang each side of trim   1-1/2 in.

> total width   _____ in.

Step 3 Measure and cut the sill to the proper width and length using the data accumulated in steps 1 and 2. Plane the sawn edge with the jack plane.

Step 4 Hold sill (flat) to the window, equalizing the overhang on each side. Mark the sill where the edge of the wall or window jamb touches the sill.

Step 5 Set the combination square for depth by holding the frame of the square to the wall and slide the blade to *just* touch the window sash or aluminum frame and tighten the screw.

Step 6 Place the frame of the square against the back side of the sill and draw a line at the end of the blade while dragging the square from the end of the sill to the mark made in step 4. Draw a line from the mark to the line just made.

Step 7 Cut the stock away from the sill in the area marked. Cut the sill so that the line is left.

Step 8 Repeat steps 6 and 7 at the opposite end of the sill.

Step 9 Install the sill in the opening and check for fit at the window or frame. If should fit gently against either.

Step 10 Nail the sill in place by face-nailing at each end within the jamb area and putting one or two nails through the center. Set the nails.

Step 11 Cut a piece of window trim equal to the length of the sill minus 1-1/2 in. for use as an apron.

Step 12 Taper the ends of the trim to the pattern of the trim (see Figure 19-12).

Step 13 Position the apron under the sill and equalize the overhang of sill on either end.

Step 14 Nail the sill to the wall with a nail near each end and one or more through the center. Set the nails.

IT7: BOXING METAL WINDOW OPENINGS

RESOURCES

Materials:

1 × _____ × _____ no. 1 or no. 2 common pine or fir
    width   lineal ft

  (select one) per window

1/2 lb 6d finishing nails per window (8d nails may be substituted)

Assorted blocking material as required

Tools:

1 6-ft folding ruler

1 13-oz claw hammer

1 no. 10 crosscut handsaw

1 no. 5-1/2 ripsaw or power saw

1 combination square

1 1/16-in. nail set

1 sawhorse

1 12-in. jack plane

ESTIMATED MANHOURS

1 hour per window

PROCEDURE

Preliminary step Install window sill as in Routine IT6.

Step 1 Rip and plane the stock to the width required:

a. Measure from the window frame to the finished side of the wall surface _____ width.

b. Measure the perimeter of the window area

2 × height _____ ft and in.

1 × head _____ ft and in.

+ 12 in. waste _____ 1 ft

total length for window ___ ft and in.

Step 2 To be sure that the boxing stock is installed at the correct place, measure and calculate the thickness of the blocks that will be needed to fill the area from the studs or header.

a. Measure from the stud or head to a point on the window frame free of window restraints and record the measurement_____.

b. Subtract the thickness of the boxing stock_____.

c. The total thickness of blocking = _____.

Step 3 Repeat step 2 at least three times on the head and sides, once near the corners, and once in the center.

Step 4 Rip blocks of 2 x 4 to the required thickness; cut for length and nail perpendicular to the run of stud or head. (*Note:* The blocks will probably be of different thicknesses because of irregularities and out-of-plumb problems.)

Step 5 Measure the full length of the head area and cut boxing stock prepared with step 1.

Step 6 Nail material to the head area by nailing through the blocking into the header. Drive two nails per blocking area. Set the nails.

Step 7 Measure the side pieces from the sill to the head and cut for a *tight* fit. (*Note:* Measure and cut each side piece separately.)

Step 8 Nail the pieces in place and set the nails.

IT8: INSTALLING CASING ON A WINDOW

RESOURCES

Materials:
_____ lineal ft casing _____ and _____. The number
   no.   type   size

equals
2 × window height + 6 in. _____
+ 1 × window width + 12 in. _____
total lineal feet per window _____.
Average 1/2 lb 6d finishing nails per window

Tools:
1 6-ft folding ruler
1 combination square
1 13-oz claw hammer
1 no. 10 crosscut handsaw
1 1/16-in. nail set
1 pair sawhorses or workbench
1 miter box (optional)

ESTIMATED MANHOURS

1 hour per window

PROCEDURE

Preliminary step: Determine whether the trim will be flush with the window jamb/boxing or recessed 1/4 in. from the edge.

Step. 1 For recessed trim only. Set the combination square for 1/4 in. Hold the body of the square along the flat surface of the jamb/boxing and scribe a line on the edge of the jamb/boxing while holding a pencil to the end of the square's blade and *sliding* the square on the jamb/boxing.

Step 2   Square one end of a piece of casing.

Step 3   Measure the length of a side piece of casing from the sill to the inside edge of the header or the 1/4-in. recess line drawn on the header. Record the length on the casing's inner edge.

Step 4   Place the combination square (used as a miter) so that its blade aligns with the mark made in step 3. Draw a 45-degree line.

Step 5   With the handsaw, cut a miter with a 2- to 3-degree undercut.

Step 6   Install the casing by nailing it into the jamb and wall (opposites) 3 in. up from the sill 8 in. OC above the bottom nails and 3 in. down from the 45-degree cut.

Step 7   Mark a 45-degree line on a second piece of casing so that when cut, it completes the first corner. Cut a miter.

Step 8   Try the piece for fit.

Step 9   Hold the casing in position and mark the uncut end for length on the inner edge.

Step 10   Mark a 45-degree miter on the casing from the inner-edge mark to the outer edge, and cut with the saw. Set aside.

Step 11   Repeat steps 2 through 5 for the opposite side piece.

Step 12   Tack-nail the side piece and head piece of the casing in place, and carefully examine the joints. If good, complete nailing and set all nails; or remove the head piece, apply glue to the miter surfaces, and reinstall.

IT9: INSTALLING DOOR AND WINDOW STOPS

RESOURCES

Materials:

_____ lineal ft (1-1/8 in.) (1-1/4 in.) (1-1/2 in.) × (3/8 in.)
    no.

(1/2 in.)

Door stop. Where 17 lineal ft provides stops for doors up to 36 in. wide.

2 × window height _____
+ 1 × window width ___
+ 6 in. for waste   6   in.
                    ‾‾‾‾‾
   total _____ ft and in.

1/4 lb. 4d finishing nails per door or window unit

Tools:
1 6-ft folding ruler
1 no. 10 crosscut handsaw
1 13-oz claw hammer
1 miter box
1 coping saw
1 1/16-in. nail set
1 sawhorse or workbench

ESTIMATED MANHOURS

30 minutes per unit

PROCEDURE

Step 1 Measure the length of the head piece from jamb to jamb.

Step 2 Cut both ends of a stop square, equal in length to the measurement in step 1.

Step 3 Position on the head jamb with the square side against the door/window sash.

Step 4 Start a 4d finishing nail approximately 2 to 3 in. from each end of the stop.

Step 5 Tack nails through the stop into the jamb. Then make the following test.

a. Check the door to see that the hinged side clears the stop when the door is opened.

b. Check that the door's leading edge on the exterior side is *flush* with the jamb's edge.

c. Check that the window sash clears the top easily. (*Note:* If the window sash does not reach to the head, measure the thickness of the sash, position, and tack the stop accordingly.)

Step 6 Measure the length from sill to head of a side piece.

Step   7   Cut a stop equal to the length measured in step 6 *with* an *open* 45-degree cut at one end.

Step   8   Cope the 45-degree cut. (See Routine IT1, if necessary.)

Step   9   Try for fit, set in position, or use as a pattern for the opposite member of the pair.

Step 10   Verify the length of the opposite side. Cut a mate to the first to make a *pair*, using steps 6, 7, and 8.

Step 11   Position the stop for nailing according to the following:

a. Door-hinge side: allow 1/16-in. clearance between the edge of the stop and the face of the door style with the door closed. Nail in place.

b. Door-face (lock) edge: Position the stop at the head to match the head stop, and nail. Close the door (assume that the lock is installed) and position the stop evenly against the door. Nail, and set all nails.

c. Window sides: Position the stop gently against the sash with the sash closed. Nail near the sill. Raise the window, reposition the side and head stops, and nail. Set all nails.

20

FLOOR INSTALLATION

Center Line in tile floor layout, a chalk line parallel to the walls through the center of the room.
Underlayment plywood, particle board, or pressed board in sheet form used under resilient flooring.

Vapor barrier 15-lb felt paper, pervious-type paper, or polyurethane sheet goods that insulate against moisture.

Flooring in the home or office is applied over a subfloor or over a concrete floor. In repair/replacement situations, the new floor is sometimes installed over the old floor.

This chapter considers installation and characteristics of two distinctly different types of flooring materials. They are *wood-strip flooring*, and *tile block, vinyl,* or *asphalt* type. A third type of material, *parquet,* is made from wooden blocks 3/8 in. thick and approximately 12 x 12 in. square and is installed similar to tile block.

SCOPE OF THE WORK

Flooring tasks are considered finish work. Therefore, attention to detail is of major importance. To do a very good job requires dedicating considerable time to planning the work. The method selected for the installation, especially on tile floors, is also reflected in the finished job. Border tiles should be equalized for size.

In almost all installations, one man can do all phases of the job. He can prepare the subfloor surface area, remove unwanted obstacles, and do layout work as well as the actual installation.

The two types of flooring characterized in this section are wood-strip flooring and tile-block flooring. As you will learn, each has a separate and unique set of conditions that must be met for satisfactory installation.

WOOD-STRIP FLOORING

Wood-strip flooring is available in hardwoods—oak, birch, and pecan—and in soft woods—fir and pine. Table 20-1 provides data that will aid you in selecting the type and quality for the job. In addition, the table provides the overall *finished* dimensions of the available sizes. You can, for instance, determine what the nominal width of the stock will be and order accordingly. A 2-3/8 in.-wide strip is nominally a 3-in. strip.

Softwood flooring costs less than most types of hardwood and may be used to good advantage in bedrooms and closets and for outside use. Hardwood flooring costs more and provides a more durable surface. The two types have similar design characteristics.

Figure 20-1 shows the characteristics of strip flooring. Figure 20-1A indicates tongue and groove with *no* end groove, tongue and groove four sides: one end and side with a tongue and the other end and side with a groove, and hollow (grooved) back for better fit with the subfloor. Figure 20-1B shows a standard design for thin flooring strips, and C shows square-edged flooring, a type that is face-nailed or pegged.

The most widely used pattern is a 25/32 in. x 2-1/4 in. strip flooring.

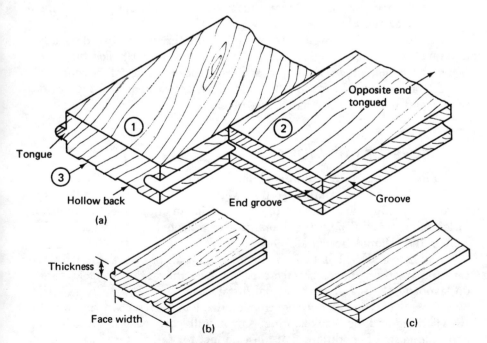

Figure 20-1 Wood-Strip Flooring

TABLE 20-1: Grade and Description of Strip Flooring

Species	Grain Orientation	Size Thickness (in.)	Width (in.)	First Grade	Second Grade	Third Grade
Softwoods						
Douglas fir and hemlock	Edge grain	25/32	2-3/8–5-3/16	B and better	C	D
	Flat grain	25/32	2-3/8–5-3/16	C and better	D	—
Southern pine	Edge grain and flat grain	5/16–1-5/16	1-3/4–5-7/16	B and better	C and better	D (and No. 2)
Hardwoods						
Oak	Edge grain	25/32	1-1/2–3-1/4	Clear	Select	—
	Flat grain	3/8	1-1/2, 2	Clear	Select	No. 1 Common
		1/2	1-1/2, 2			
Beech, birch, maple, and pecan		25/32	1-1/2–3-1/4	First grade	Second grade	—
		3/8	1-1/2, 2			
		1/2	1-1/2, 2			

The strips are laid lengthwise in a room and normally at right angles to the floor joists. If strip flooring is installed over sheathing boards, some type of vapor barrier must be installed over the subfloor *before* the flooring is installed.

This type of flooring is customarily laid *before* the door units or jambs, baseboard, and shoe molding are installed. To do the job well requires a very critical inspection of the subfloor surface. The floor should be swept clean and all nails should be partially set. If underlayment plywood panels are used, all cracks must be even and free of protrusions.

Figure 20-2 shows how the first row of flooring should be installed. The strip of flooring is placed 1/2 to 5/8 in. away from the wall. The space is needed to allow for expansion of the floor when the moisture content increases. An 8d finishing nail should be driven through the flooring near the wall and into the subfloor and floor joist. This nail will be covered with the baseboard and shoe molding. In addition, an 8d flooring nail should be driven into the tongue area at joist centering distances to hold the tongue side to the subfloor.

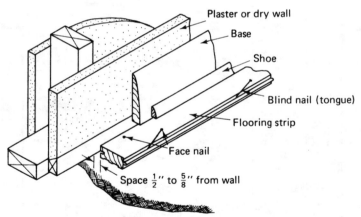

Figure 20-2 Installing the First Strip

Usually more than one piece of the flooring is needed per row; therefore, complete the first row before starting the second row. An important point to remember is to use the remainder of the last installed piece as the beginning piece of the second row. This technique reduces waste and almost always simplifies breaking the joint.

Refer to Figure 20-3 and study the details provided. In A the significant factor is the angle at which the toe-nailing is done. B shows what the floor will look like if hammer marks are made. To prevent this from happening, use your nail set as shown. Also, although not shown, try to

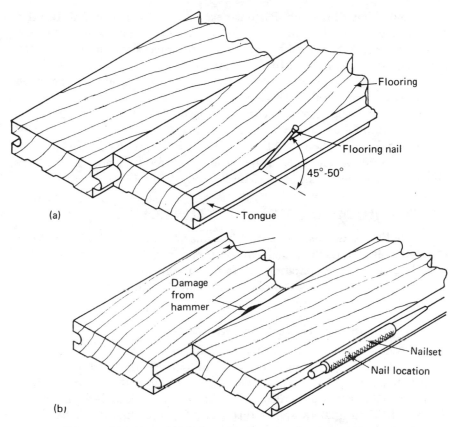

Figure 20-3 Proper Nailing Technique for Strip Flooring

avoid splitting or caving the tongue while nailing. If either does happen, use the claw of your hammer and realign the stock or remove the damaged section. There is only one place for all the tongue to go, and if it does not fit into the groove, spacing between the boards results.

The last piece or pieces that will fit against the wall must be ripped so that when installed there will be 1/2- to 5/8-inch clearance between the board and the wall. Toe nailing is not possible; top nailing must be used.

Tips for Wood-Floor Installation

The last two rows near a wall may be difficult to nail and draw tightly. The next-to-last row must be toe-nailed and the last row must be top-nailed. To draw the board tightly for the next-to-last row, cut a scrap of flooring about 12 in. long. Place it next to where you will nail. Insert a flat bar or crow bar behind the scrap and another scrap behind the bar, making a lever. Apply pressure and when the pieces are drawn, nail the flooring. Use the same idea for "snugging up" the last course of flooring.

During the laying of the floor, some pieces may provide a challenge to drawing them tight. If this occurs, cut a scrap piece approximately 14 in. long and slip it over the tongue. Drive two flooring nails simultaneously into the tongue area of the scrap piece. After starting the nails, lift the block off the floor 3/8 to 1/2 in. and drive the nails into the subfloor. As you drive the nails in, really let the force flow and don't worry about marring your scrap block. The flooring strip *will* move in place.

TILE BLOCK FOR FLOOR INSTALLATIONS

Tile blocks are sold in 12 x 12 in. squares and some 9 x 9 in. blocks may be available. Each is inlaid in the same way. The basic segments of the job include the preparation of the underlayment, laying out the area, applying the cement, and cutting and fitting the tiles and border tiles. We shall examine each segment in order.

Preparation of Underlayment

Resilient floors should *not* be installed directly over a board or plank subfloor. Underlayment-grade plywood or particle board should be installed. Its thickness should be selectively determined. If, for example, the entire floor area will be tiled, a 1/4- or 3/8-in.-thick underlayment will do the job. Figure 20-4A shows this view; B shows the alternative. The underlayment's thickness, 5/8 in. added to the tile's thickness, 1/8 in. equals the thickness of the strip flooring.

Concrete floors that will be tiled need a smooth, polished finish. A vapor barrier must be installed under the concrete prior to pouring so that the moisture is kept from disturbing the tile cement and bond.

Where a new tile floor is installed over an old tile floor the shoe molding around the room must first be removed. Next, a close examination of the surface must be made. Loose tile (old) must be replaced and cracks filled and allowed to harden; and a thorough cleaning and stripping of the floor is needed. In most installations the new tile may be cemented directly to the old. However, the joints should be offset by 1 in. This will unbalance the border tiles but only minimally.

Laying Out the Area

Two points should be considered during the layout. The longest run in the room should be parallel with the wall borders rather than the short run. Irregularities, out of square, and unusual wall arrangements

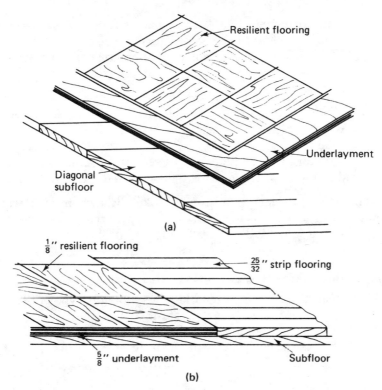

Figure 20-4 Selecting the Underlayment's Thickness

are usually minimized by this technique. Also, the border tiles on all four walls should equal one-half or more the width of a tile.

The formulas provided here will establish an even border-tile width and a basis for calculating the total number of tiles needed.

1. Room width
   a. Measure the room width and divide by 2.
   b. Divide the result by the width of the tile (9 or 12 in.) and
   c. If the remainder (in inches) is 3/4 in. to one-half the width of the tile, *offset the center tile* by one-half its width, or
   d. If the remainder (in inches) is one-half or more or the width of the tile, use the center line as the starting edge.

Figure 20-5 shows both arrangements.

2. Room length
   a. Measure the room's length and divide by 2.
   b. Divide the results by the width of the tile (9 or 12 in.).

    c. If the remainder is 3/4 in. to one-half the width of the tile, *offset the center tile* by one-half its width, or

    d. If the remainder is one-half or more than the width of the tile, use the center line as the starting edge.

3. Calculate number of tiles needed
    a. Multiply step b's results (whole tiles) by 2 _____
    b. Add 2   _____
    c. Total for room width   _____
    d. Total for room length   _____
    e. Multiply steps c and d for total required _____

Using your calculations measure a like distance from a wall at each end of the room and snap a chalk line. Repeat the process the opposite way, creating four blocks (refer to Figure 20-5).

Applying Cement and Cutting and Fitting Tiles

Apply cement for tile (type and kind according to type of tile being laid) over a 4- to 8-sq ft area in one quadrant. When set (dried), properly lay tile onto the cement along the chalk line previously snapped. When possible, start at the center where the chalk lines cross. Install the tiles,

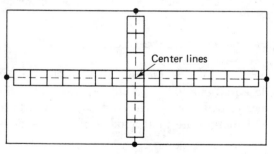

Step *C* of calculations

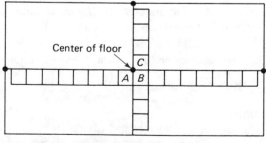

Step *D* of calculations

Figure 20-5 Layout for Tiles

butting the edges gently, until all the cemented surface is covered. Repeat the process up to the border tile.

Border-Tile Cutting and Fitting

The easiest, quickest, and most economical method to fit the border tile is shown in Figure 20-6. Align a border tile, evenly overlaid, atop the last tile installed. Take another tile from the box and position it against the wall and overlay on the border tile. Scribe the border tile along the line provided by the piece butted to the wall. Remove both pieces of tile and cut the border tile. Fit in place and press down into the cement.

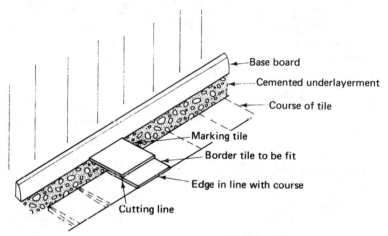

Figure 20-6 Laying Out, Fitting, and Marking Tile for Borders

Cuts around the door casing and utilities, where they occur, must be made on an individual basis. Where possible, work from the last course of tile installed rather than a wall edge.

To finish the job, a shoe molding must be installed along the baseboard. See Routine IT4, Chapter 19, for aids to installing the shoe molding.

IF1: INSTALLING STRIP FLOORING

RESOURCES

Materials:

_____ square ft flooring by type and grade
　no.

type _____ (tongue and groove 4 sides, 2 sides)

grade _____ No. 1 clear, select, common

kind _____ oak, fir, pine

100 bd ft of tongue-and-groove flooring covers approximately 67 sq ft of floor

10 percent for waste

_____ lb flooring nails (5 lb per 100 sq ft)

Tools:

1 16-oz hammer or special floor hammer

1 6-ft folding ruler

1 no. 8 crosscut handsaw

1 flat 3-in. pry bar (cold chisel)

1 pair sawhorses

1 power saw (optional for ripping)

1 chalk line

1 crow bar

1 combination square

ESTIMATED MANHOURS

4 hours per 80 sq ft

PROCEDURE

Step 1   Inspect the floor area and remove any object that will prevent the floor from laying flat.

Step 2   If required, cut and install a strip of flooring perpendicular to the floor joist run with the side groove against the wall, the end groove against the corner, and toe-nailed at 12- to 16-in. intervals.

Step 3   Measure and cut a second or last flooring strip to complete the first row. *Be sure to* cut the end with a tongue (if tongue-and-groove 4 sides is used).

Step 4   Slide the cut piece into the end tongue and against the wall. Draw the end joint tight by using either a crowbar or a flat bar at the opposite end of piece. Toe-nail to the subfloor.

Step 5   Nail the remainder of the piece cut in step 3 as the first piece of the second row.

Step 6   Repeat steps 3, 4, and 5 on a row-by-row basis across the room.

Caution:
1. *Do not* strike the top edge of the flooring with a hammer. Hammer marks will always show.
2. Use a nail set or a second floor nail on its side to drive the toe nail home.
3. Check your work at 4-ft intervals for straightness and parallel to the opposite wall.

IF2: INSTALLING TILE-BLOCK FLOORING

RESOURCES

Materials:

_____ square ft of floor space. Each tile is 12 × 12 in. See the
   no.

layout for data to determine the number of square feet.

_____ cement
  gallons

Tools:
1 6-ft folding ruler
1 utility knife
1 torch (for asphalt tile only)
1 trowel (grooved)
1 chalk line
1 framing square
1 workbench

ESTIMATED MANHOURS

4 hours per 100 sq ft

PROCEDURE

Step 1 Inspect the subfloor or old floor for defects. Then

a. Install underlayment (as required).
b. Spackle cracks and crevices.
c. Smooth protrusions and sand rough spots smooth.
d. Wash with detergent.
e. Drive all nails slightly (1/16 in.) below the floor surface (as appropriate).
f. Remove the shoe molding (as required).

Step 2 Lay out the floor area for the snapping of chalk lines.
a. Measure a like distance from a long wall at both ends of the room according to the data in the chapter.
b. Snap a line connecting the two points.
c. Measure a like distance from a short wall at both ends of the room according to the data in the chapter.
d. Snap a line connecting the two points.

Step 3 Apply cement in one quadrant covering an area approximating 6 to 12 sq ft. Allow to dry.

Step 4 Place the first tile at the intersection of the chalk lines and into the cement.

Step 5 Install all the tile over the cemented area.

Step 6 Repeat steps 3, 4, and 5 until all but the border tiles are laid.

Step 7 Lay a border tile exactly over a tile adjacent to the border, and with another tile butted to the wall, scribe the border tile along the line provided (see Figure 20-6). Cut the border tile along the line scribed. Set in place and into the cement.

Step 8 Corner tiles may be fitted by using step 7 first for one wall, then for the adjacent wall.

21

PANELING

BASIC TERMS

Adhesive specially designed glue that will secure a panel to wood studs or gypsum panels.
Furring applying a strip of wood to a wall to prepare it for paneling, or to a jamb to fill out the jamb to wall-surface thickness.
True trade term that means a surface is straight throughout its entire length.

Paneling walls in a room is frequently associated with adding rooms to homes or remodeling rooms that already have some type of wall covering. Data and aids for both types of installations are provided.

SCOPE OF THE WORK

Panel material is a prefinished product; therefore, any carelessness or mistakes result in a bad job and additional expense. In this task *being right* is much more important than *being fast*. Dedicate plenty of time to planning, laying out panels for cutting, and to installation. Program your work so that it runs smoothly and efficiently. Doublecheck all measurements *before* cutting.

Usually one person can install paneling, but a helper will make the work easier. A low bench or stepladder will be needed to fasten panels near the ceiling. A cutting surface such as a pair of sawhorses and two 2 x 4 boards 8 ft long will make a comfortable, convenient cutting area.

Next we shall discuss general characteristics of paneling and fastening techniques: layout, cutting, and fitting; installing corner bead; and problems to overcome when remodeling with paneling.

CHARACTERISTICS OF PANELING

The standard configuration of panel is shown in Figure 21-1. There is a distinct pattern to the V grooves in the panel. One side starts with a

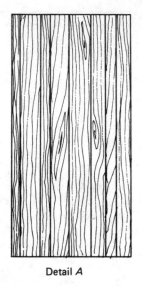 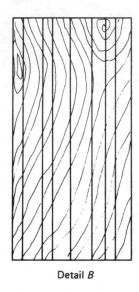

Detail *A*   Detail *B*

Figure 21-1 Paneling

narrow panel, then groove; the other side starts with a wide panel, then groove.

The graining patterns of the pieces in Figure 21-1A show that the veneer is laid in strips no wider than 12 in.; in B, the grain pattern runs the full 4-ft width. Each style creates a totally different appearance. No absolute rule is established as to which to use, but the narrow-strip veneer type is usually more pleasing in small rooms.

Two requirements need to be met when nailing paneling: edge nailing and intermediate nailing (Figure 21-2). Nails, colored the same as the panel and long enough to go through the panel and 3/4 in. into the stud, must be nailed 4 to 6 in. OC along the edges. Nails along the edge of the second and subsequent panels should be opposite those of the previous panel. Nails driven in intermediate (16-in.-OC framing) studs or furring should be nailed at 8 in. OC.

Adhesive for gluing panels to studs or gypsum should be purchased in tubes, and applied with a caulking gun. Where the edges of panels will be fastened, a solid bead of adhesive should be applied to the member or gypsum (Figure 21-3). On intermediate members and 16 in. OC, an intermittent bead should be applied, with 3-in. beads and 6-in. spaces.

Some types of adhesive are made to allow the panel to be pressed in place and allowed to dry. Others require that the panel be pressed in place and then partially removed—that is, remove only far enough to allow the adhesive to set. The best way to do this is to tack-nail the panel

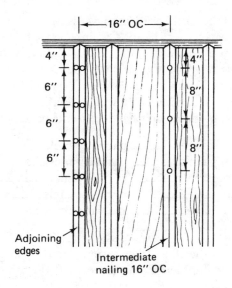

Figure 21-2 Nailing Pattern for Paneling

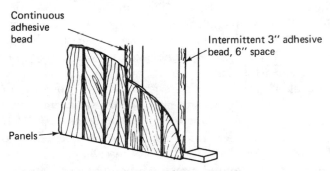

Figure 21-3 Gluing Panels

near the ceiling and prop the base open with a 6-in. block of wood. Routine IP3 provides step-by-step procedures. Once the adhesive has set, approximately 10 to 15 minutes, the panel is pressed into place.

LAYOUT, CUTTING, AND FITTING

The rule to follow for layout is that panel edges must break on the center of a stud, except at corners. A panel is 48 in. wide, so your layout must locate a common stud up to 48 in. from a corner. Thereafter there must be another common stud each 48 in. Between these critically spaced studs, others must be located 16 in. OC. If the studding is 24 in. OC. install blocking (cats) each 16 in. OC in horizontal rows from floor to ceiling, or apply gypsum panel (3/8 in. thick) to the stud wall.

To determine how many panels will be needed, proceed as follows. Measure each wall's length and height and record the measurements on a sketch of the room. Divide each wall's length by 4 to determine the number of full and partial panels needed in inches. List them accordingly.

Wall A: no. panels_____ partial panels_____ in.
Wall B: no. panels_____ partial panels_____ in.
Wall C: no. panels_____ partial panels_____ in.
Wall D: no. panels_____ partial panels_____ in.
Wall E: no. panels_____ partial panels_____ in.
   subtotal   _____

From your survey of the 48-in. placement of studs, study the partial-panel requirements and determine if one panel will do for two requirements. Add the number of panels you determine it will take to account for the pieces to the subtotal of full panels and order the grand total.

grand total_____

Two routines are provided for cutting and fitting panels. They are IP4, Fitting Corners for Panels, and IP5, Cutting Out for Windows, Doors, and Utilities. There is a singleness of purpose in fitting corners. When done properly the joint is accurately fit and is free of molding. If poorly done, a piece of 3/4-in. quarter-round molding will need to be installed. There is commonality in cutting out for window, door, and utility openings on a panel. Therefore, these are grouped under one routine.

A corner is perfectly straight or it is not. By placing a straightedge in the corner and examining how well it fits, you can easily determine its accuracy. If it is straight or "true," measurements taken from the corner to the edges of the last panel installed at the ceiling and at the floor will provide all the fitting required, provided the cut is accurate. Transfer the width to a panel, snap a chalk line, and cut along the line.

Where the corner is not true, special fitting is needed. Figure 21-4 shows some of the elements. First, measure the width of the piece to be fitted. Add 1-1/2 in. to this width and rip a panel accordingly. Next measure *back* 2 in. from the edges of the last panel installed at the ceiling and floor levels. Position the manufactured edge of the panel just ripped, along the 2-in. marks, with the top end of the panel toward the ceiling, and tack-nail in place. Cut a 2 x 2 in. block of wood. While holding it against the corner and the pencil against the block, slide the block and the pencil from the ceiling to the floor, at the same time marking a pencil line on the panel. Remove the panel to the bench, and follow the pencil

Figure 21-4 Fitting Panels to an Irregular Corner

line with your saw very, very accurately. Reposition the piece and nail or glue in place.

Figure 21-5 illustrates the methods used for cutting out for windows, doors, and utilities. The letters showing details correspond to instructions in Routine IP5.

Ceiling line/reference line

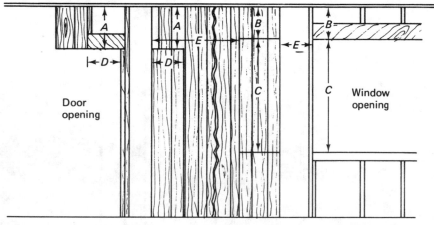

Figure 21-5 Cutting Out for Windows, Doors, and Utilities

INSTALLING OUTSIDE CORNER BEAD

If you elect to use an outside corner bead (Figure 21-6) you will eliminate the need for making a very difficult 45-degree miter 8 ft long. In addition, the bead affords some added protection for corners, which usually take a physical beating. The bead is always nailed or glued in place *after* the baseboard and ceiling trim are installed.

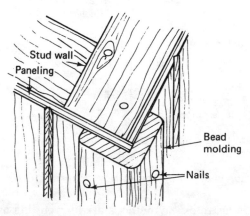

Figure 21-6 Corner-Bead Installation

PROBLEMS WHEN REMODELING WITH PANELING

To prepare wall surfaces to be paneled, a careful inspection of the wall must be made. In addition, all door, window, and wall trim and molding must be removed.

The inspection should consider the following set of items.

INSPECTION LIST	YES/NA	NO
1. Present wall surface is sound	_____	_____
2. Loose areas have been secured	_____	_____
3. Stud placement and centering have been determined	_____	_____
4. Door jamb/casing or just casing is removed	_____	_____
5. Baseboard and ceiling molding are removed	_____	_____
6. Furring strips 16 in. OC are installed	_____	_____
7. Wall insulation and soundproofing is installed	_____	_____

If interior door units are installed, carefully drive any finishing nails through the jamb and gently pry one-half of the jamb and casing free from the wall.

If an exterior or built-up jamb is installed, gently pry the casing from the jamb. Try to save the trim, for two reasons: (1) new trim will not exactly match the old, and (2) using the old trim will save money. Remove the casing and apron from the window units. Do it gently to save the material. (*Tip:* To remove the nails from trim taken from a jamb, *do not* back the nails through the face of the trim; pull the nails through from the *back* or snip off close to the back side of the trim.)

When paneling is added to the wall, those doors and windows that have fixed jambs will require a furring strip to account for the panel's thickness. The panel's edge cannot be seen when the job is complete. Therefore, you must either cut 1/4 in. x 3/4 in. furring strips or buy 1/4 in. x 1-1/2 in. lattice strips and rip them in two. These strips will need to be cut to fit so that when glued (white glue) and nailed to the jamb's edge, they will be flush with the outside face of the jamb (Figure 21-7).

ROUTINES

Following are seven routines which provide the aids needed to do every type of paneling job. Routine IP1 describes how to fur cement walls in preparation for paneling; IP2 and IP3 outline installation techniques;

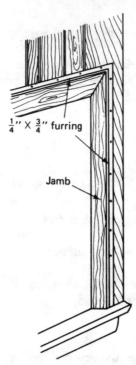

$\frac{1}{4}'' \times \frac{3}{4}''$ furring

Jamb

Figure 21-7 Furring a Jamb

IP4 and IP5 list the steps needed to cut and fit panels; IP6 describes the installation of corner bead; and IP7 tells how to fur window and door jambs.

IP1: FURRING CONCRETE WALLS FOR PANELING

RESOURCES

Materials:

_____ 1 x 2 or 1 x 3 x 8 ft furring strips. Three per 4 ft plus 1 for
   no.

each wall is needed.

_____ 1 x 2 or 1 x 3 for horizontal stripping
   no.

_____ lb 8d cement (case-hardened) nails. One pound is enough
no.

for 4 to 5 furring strips.

Tools:
1 no. 8 crosscut handsaw
1 24-in. level
1 16-oz claw hammer
1 6-ft folding ruler

ESTIMATED MANHOURS

4 hours per 8 ft of wall

PROCEDURE

Step 1   Precut sufficient furring strips for a wall to reach from the floor
to the ceiling.

Step 2   Nail a strip in each corner with nails spaced 12 in. apart.

Step 3   Lay out and mark 16-in. spacings from the left or the right
corner along the wall or the floor near the wall.

Step 4   Position a furring strip centered on the 16-in. mark and nail.
Place the level alongside the furring strip and, when plumb, nail the remainder of the piece.

Step 5   Repeat step 4 until all strips are installed. Repeat steps 2 through
5 for all walls.

Step 6   *Optional:* Cut horizontal pieces between upright strips and nail.

IP2: INSTALLING PANELING ON STUDS OR FURRING WITH NAILS

RESOURCES

Materials:
_____ of panels 4 ft x 8 ft _____
no.   type

wall 1 length _____   no. panels _____
wall 2 length _____   no. panels _____
wall 3 length _____   no. panels _____
wall 4 length _____   no. panels _____
wall 5 length _____   no. panels _____
     total   no. panels _____

Note: Sometimes portions of panels may be used on different walls.

Tools:
1 13-oz claw hammer
1 bench or stepladder
1 6-ft folding ruler
1 3-in. flat bar
1 no. 10 crosscut handsaw
1 chalk line or 8-ft straightedge

ESTIMATED MANHOURS

20 minutes per panel, excluding utility and opening cutout times

PROCEDURE

Step 1 Measure from a corner to a common stud not more than 48 in. away and determine the width to the stud's center. Verify that the distance is equal along the floor and near the ceiling.

Step 2 Using the distance measured in step 1, lay out and cut a panel to the required width. (*Caution:* the ripped edge *will* fit into the corner; the manufactured edge must break on the center of the stud.)

Step 3 Install the panel in the corner, slip the flat bar under the panel, and raise the panel to the ceiling. Nail first at eye level, with a nail at opposite edges.

Step 4 Verify the position of the panel, adjust if needed, and nail the edges 4 in. OC and intermediate studs 8 in. OC.

Step 5 Butt the next panel *gently* against the first panel's edge, raise the panel to the ceiling, and nail. (*Note:* Precut utilities window and door openings before installing.)

Step 6 Finish the installation by using Routine IP4, Fitting Corners for Panels.

Step 7 Repeat steps 1 through 6 for each wall.

IP3: INSTALLING PANELING BY GLUING TO UNDERSURFACES

RESOURCES

Materials:

_____ of 4 ft x 8 ft panels to cover the walls. See Routine IP2 to
  no.

estimate the panels required.

_____ tubes adhesive. One tube is enough for 2 to 2-1/2 panels.
  no.

A few 6d finishing nails

Tools:

1 no. 10 crosscut handsaw
1 3-in. flat bar or crowbar
1 6-ft folding ruler
1 caulking gun
1 13-oz claw hammer
1 bench or stepladder
1 chalk line or 8-ft straightedge

ESTIMATED MANHOURS

20 minutes per panel, excluding utility and opening cutout times

PROCEDURE

Step 1   Measure from a corner to a common stud not more than 48 in.
away, and determine the width to the stud's center. Verify that the distance is equal along the floor and near the ceiling.

Step 2   Using the distance measured in step 1, lay out and cut a panel to
the required width. (*Caution:* The ripped edge *will* fit into the corner;
the manufactured edge must break on the center of a stud.)

Step 3   Place the panel in position and verify the fit. Remove the panel
from the wall.

Step 4   Insert a tube of adhesive in the caulking gun. Snip the end of the tube and apply a bead of adhesive on each stud (or wall board) where the corner and edge of the panel will be. Alternate 3-in. beads and 6-in. spaces on intermediate studs or 16 in. OC.

Step 5   Place the panel against the wall, raise to the ceiling, and press into the adhesive.

Step 6   Drive 3 to 4 finishing nails through the panel near the ceiling. Then prop the bottom of the panel away until the glue sets (see the tube of adhesive for drying time).

Step 7   Remove the prop and press the sheet into the adhesive. Remove the nails or drive the nails home.

Step 8   Fit the second (and subsequent) panel, butting the edges gently. Apply adhesive and press the panel to the wall.

Note: See other routines for fitting panels and Chapter 19 for trimming instructions.

IP4: FITTING CORNERS FOR PANELS

RESOURCES

Materials:
None required (already figured in other routines)

Tools:
1 chalk line
1 13-oz claw hammer
1 7-ft straightedge
1 6-ft folding ruler
1 no. 10 crosscut handsaw
1 2-in. marking block

ESTIMATED MANHOURS

30 minutes per corner

PROCEDURE A: CUTTING FOR A TRUE CORNER

Step 1 Verify the straightness of the corner by holding a straightedge against the wall.

Step 2 Measure and record the width of the piece being fitted from the edge of the last panel installed to the wall near the floor and ceiling levels.

Step 3 Lay out the panel to be cut for a corner fit using the results of step 2.

Step 4 Connect the lines by snapping a chalk line.

Step 5 Slightly undercut the panel along the chalk line.

Step 6 Position the sheet and verify the fit. Adjust minor irregularities by smoothing with a block plane.

PROCEDURE B: CUTTING FOR AN IRREGULAR CORNER

Step 1 Measure and rip a panel 1-1/2 in. wider than the widest point of the place where the panel will be installed.

Step 2 From the edge of the last panel installed, measure back 2 in. and drive a 4d finishing nail close to the ceiling and near the floor.

Step 3 Position the factory edge of the panel ripped in step 1 against the 4d nails and tack the panel in place.

Step 4 Cut a marking/scribing block 2 in. x 2 in. x 3/4 in. thick.

Step 5 Place the block back on the tacked panel, edge against the corner, and scribe a line 2 in. out from the corner (pencil against the block) from the ceiling to the floor.

Step 6 Remove the nails from the panel, place across the sawhorses, and rip along the scribed line. Follow *all* the irregularities drawn for an accurate fit.

Step 7 Position the panel in place and nail.

IP5: CUTTING OUT FOR WINDOWS, DOORS, AND UTILITIES

RESOURCES

Materials:
None required

Tools:
1 6-ft folding ruler
1 keyhole saw
1 brace and 1-in. auger
1 48- to 60-in. straightedge
1 no. 10 crosscut handsaw

ESTIMATED MANHOURS

15 minutes per cutout

PROCEDURE

Step 1 From the ceiling, measure to the top and lower sides of the area to be cut out. Transfer the measurements to a panel (face side up).

Step 2 From a corner or an edge, or from the end of an installed panel, measure the far- and near-side distances of the cutout. Transfer the measurements to the panel.

Step 3 Connect the measurement points with a pencil and a straightedge (except the utility pipes) to form the cutout.

Step 4 Check your measurements from panel to wall.
Note: Use steps 6 and 7 for internal cuts; use steps 8 through 10 for circle cuts.

Step 5 Drill holes at the corners with the brace and auger.

Step 6 With a keyhole saw, cut along the line 4 to 5 in.

Step 7 With a handsaw, cut both lines, starting at the edge.

Step 8 Draw a circle with dividers connecting all four points measured.

Step 9 Drill a pilot hole in the waste-material area with the brace and auger.

Step 10 Cut the circle out with a keyhole saw.

IP6: INSTALLING CORNER-BEAD MOLDING

RESOURCES

Materials:
1 8-ft corner bead for each outside corner
24 brads or 3d finishing nails per corner

Tools:
1 6-ft folding ruler
1 13-oz claw hammer
1 no. 10 crosscut hand saw
1 1/16-in. nail set

ESTIMATED MANHOURS

20 minutes per corner

PROCEDURE

Step 1 Inspect the corner that is to receive molding and correct any irregularities in joining by nailing or shimming, and excessive overlap of panels by planing with the block plane.

Step 2 Measure the length of molding needed from the top of the baseboard to the underside of the ceiling molding.

Step 3 Trim one end of a corner bead square. Measure for length; mark and cut square.

Step 4 Position a strip on the corner and start a nail in both flanges of the corner molding approximately 12 in. up from the bottom. Hold the molding and drive in the nails. (*Note:* You may find that slight toe nailing gives more holding strength.)

Step 5   Drive two more nails in 3 in. up from the bottom.

Step 6   Ensure that the molding fits under the ceiling molding (slight bowing may be needed at the middle of the molding).

Step 7   Drive two nails 3 in. down from the top end and every 8 in. OC thereafter. Set all nails with the nail set.

IP7: MODIFYING WINDOW AND DOOR JAMBS

RESOURCES

Materials:
1/4 x 3/4 in. strips for each window jamb and door jamb (except door units, which can be removed as a unit). 1/4-in. lattice ripped in two is ideal.
1 lb 2d finishing nails or brads
1 pint white glue

Tools:
1 no. 8 crosscut handsaw
1 6-ft folding ruler
1 13-oz claw hammer
1 sawhorse

ESTIMATED MANHOURS

30 minutes per unit

PROCEDURE

Step 1   Prepare stock for building out jambs to 1/4 in. x 3/4 in.

Step 2   Measure and cut a strip for a side jamb. Glue the back side and nail with the edge flush to the window sash or the door side of the jamb.

Step 3   Repeat step 2 for the opposite jamb and bead.

APPENDIX A

LIST OF ROUTINES BY CATEGORY

Title/Routine	Chapter
Shingling	11
BRS1 Strip-shingle application	
BRS2 Shingling a valley	
BRS3 Shingling a ridge	
BRS4 Flashing a lean-to roof	
Siding	16
BSid1 Laying out siding	
BSid2 Installing building paper	
BSid3 Installing corner boards	
BSid4 Installing starter strips	
BSid5 Installing flashing over doors and windows	
BSid6 Installing lapsiding	
BSid7 Making a self-corner on lapsiding	
BSid8 Cutting, nailing, and installing asbestos siding	
BSid9 Cutting, nailing, and installing shingles	
BSid10 Cutting, nailing, and installing vertical tongue-and-groove	
BSid11 Installing vertical-panel siding	
BSid12 Caulking	
Trim	19
IT1 Installing coping molding for interior corners	
IT2 Installing door casing/trim	
IT3 Installing baseboard	
IT4 Installing shoe molding	
IT5 Installing ceiling-border molding	
IT6 Installing window sill and apron	
IT7 Boxing metal window openings	
IT8 Installing casing on a window	
IT9 Installing door and window stops	
Wall Framing	7
BF1 Wall layout and construction	
BF2 Preparing outside corner posts	
BF3 Constructing a window-frame unit	
BF4 Constructing a door-wall unit	
BF5 Constructing inside corners	
BF6 Erecting and plumbing walls	
BF7 Installing corner bracing	
BF8 Double-plate installation	
Window-Unit Installation	13
BW1 Installing Aluminum and wooden type A window units	
BW2 Installing wooden type B window units	

APPENDIX B

TOOL KIT

The following tool kit constitutes an adequate supply and variety of tools with which to perform the tasks described in this book.

All illustrations have been provided by courtesy of Stanley Tools.

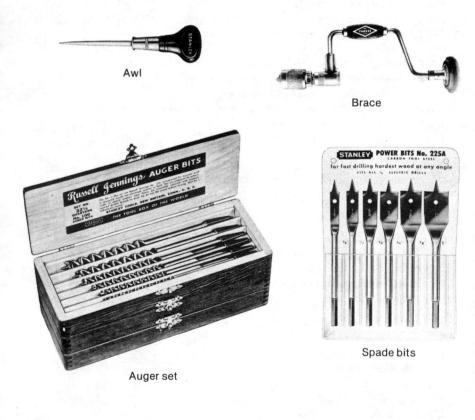

Awl

Brace

Auger set

Spade bits

Expansion bit

Putty knife

Nail set

Crowbar

Pry bar

Chisel set

Chalk line

Utility knife

Hatchet

Hacksaw

Miter box with saw

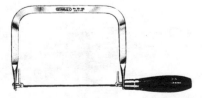

Coping saw

Handsaw

Keyhole saw

Plumb bob

Hammer

Block plane

Jack plane

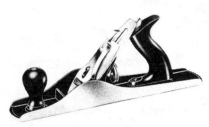

Smoothing plane

Rabbet plane

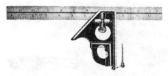

Combination square

Ruler

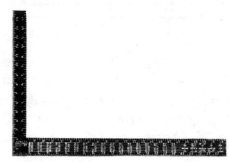

Framing square

Bevel square

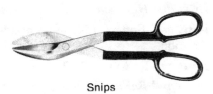

Snips

Yankeer screwdriver

Yankeer screwdriver

Common-point screwdriver

Cross-point or
phillips-head screwdriver

Level

Clamp/gripper

REFERENCE LIST

Builder 3 and 2. Bureau of Naval Personnel.

Building Guide, MOD 24. American Plywood Association.

Ceiling Tile. Armstrong World Industries.

Cement, Mason's Guide. Portland Cement Association.

Concrete Improvements Around the Home. Portland Cement Association.

Drywall Construction Handbook. United States Gypsum Corp.

Exteriors, Hardboard Siding. Masonite Corporation.

Joists and Rafters, Maximum Spans. Southern Forest Products Association.

Joists and Rafters, Span Tables for. Western Wood Products Association.

Joists and Rafters, Working Stresses for. Western Wood Products Association.

Lumber, General Information. Western Wood Products Association.

Paneling, How to Install G-P Factory-Finished. Georgia-Pacific Corp.

Plywood, Guide to Grades. American Plywood Association.

Plywood, The "How to" Book. American Plywood Association.

Sheetrock, How to Install and Finish. United States Gypsum Corp.

Wood Door Units, Hinged Interior. U.S. Department of Commerce.

Wood-Frame House Construction. U.S. Department of Agriculture.

Wood Homes for Rural America, Low Cost, Construction Manual. U.S. Department of Agriculture.

INDEX

Other Practical References

Carpentry Estimating
Simple, clear instructions show you how to take off quantities and figure costs for all rough and finish carpentry. Shows how much overhead and profit to include, how to convert piece prices to MBF prices or linear foot prices, and how to use the tables included to quickly estimate manhours. All carpentry is covered: floor joists, exterior and interior walls and finishes, ceiling joists and rafters, stairs, trim, windows, doors, and much more. Includes sample forms, checklists, and the author's factor worksheets to save you time and help prevent errors. **320 pages, 8½ x 11, $25.50**

Roof Framing
Frame any type of roof in common use today, even if you've never framed a roof before. Shows how to use a pocket calculator to figure any common, hip, valley, and jack rafter length in seconds. Over 400 illustrations take you through every measurement and every cut on each type of roof: gable, hip, Dutch, Tudor, gambrel, shed, gazebo and more. **480 pages, 5½ x 8½, $19.50**

Contractor's Guide to the Building Code
Explains in plain English exactly what the Uniform Building Code requires and shows how to design and construct residential and light commercial buildings that will pass inspection the first time. Suggests how to work with the inspector to minimize construction costs, what common building short cuts are likely to be cited, and where exceptions are granted. **312 pages, 5½ x 8½, $16.25**

Rough Carpentry
All rough carpentry is covered in detail: sills, girders, columns, joists, sheathing, ceiling, roof and wall framing, roof trusses, dormers, bay windows, furring and grounds, stairs and insulation. Many of the 24 chapters explain practical code approved methods for saving lumber and time without sacrificing quality. Chapters on columns, headers, rafters, joists and girders show how to use simple engineering principles to select the right lumber dimension for whatever species and grade you are using. **288 pages, 8½ x 11, $14.50**

Finish Carpentry
The time-saving methods and proven shortcuts you need to do first class finish work on any job: cornices and rakes, gutters and downspouts, wood shingle roofing, asphalt, asbestos and built-up roofing, prefabricated windows, door bucks and frames, door trim, siding, wallboard, lath and plaster, stairs and railings, cabinets, joinery, and wood flooring. **192 pages, 8½ x 11, $10.50**

Wood-Frame House Construction
From the layout of the outer walls, excavation and formwork, to finish carpentry, and painting, every step of construction is covered in detail with clear illustrations and explanations. Everything the builder needs to know about framing, roofing, siding, insulation and vapor barrier, interior finishing, floor coverings, and stairs. . . complete step by step "how to" information on what goes into building a frame house. **240 pages, 8½ x 11, $11.25. Revised edition**

Stair Builders Handbook
If you know the floor to floor rise, this handbook will give you everything else: the number and dimension of treads and risers, the total run, the correct well hole opening, the angle of incline, the quantity of materials and settings for your framing square for over 3,500 code approved rise and run combinations—several for every 1/8 inch interval from a 3 foot to a 12 foot floor to floor rise. **416 pages, 8½ x 5½, $12.75**

Rafter Length Manual
Complete rafter length tables and the "how to" of roof framing. Shows how to use the tables to find the actual length of common, hip, valley and jack rafters. Shows how to measure, mark, cut and erect the rafters, find the drop of the hip, shorten jack rafters, mark the ridge and much more. Has the tables, explanations and illustrations every professional roof framer needs. **369 pages, 8½ x 5½, $12.25**

Handbook of Construction Contracting Vol. 1 & 2
Volume 1: Everything you need to know to start and run your construction business; the pros and cons of each type of contracting, the records you'll need to keep, and how to read and understand house plans and specs to find any problems before the actual work begins. All aspects of construction are covered in detail, including all-weather wood foundations, practical math for the jobsite, and elementary surveying. **416 pages, 8½ x 11, $21.75**

Volume 2: Everything you need to know to keep your construction business profitable; different methods of estimating, keeping and controlling costs, estimating excavation, concrete, masonry, rough carpentry, roof covering, insulation, doors and windows, exterior finish, specialty finishes, scheduling work flow, managing workers, advertising and sales, spec building and land development and selecting the best legal structure for your business. **320 pages, 8½ x 11, $24.75**